ROLLING THROUGH AFGHANISTAN

LIFE AS A COMBAT MEDIC ENDURING PTSD

NICHOLAS BERTUCCI
HOSPITAL CORPSMAN 2ND CLASS USN
FLEET MARINE FORCE

ROLLING THROUGH AFGHANISTAN
© Copyright 2021 Nicholas Bertucci

First print publication: December, 2021

All rights reserved. No part of this publication may be reproduced, distributed or transmitted in any form or by any means, including photocopying, recording, or other electronic or mechanical methods, without the prior written permission of the publisher, except in the case of brief quotations embodied in critical reviews and certain other noncommercial uses permitted by copyright law.

Although the author and publisher have made every effort to ensure that the information in this book was correct at press time, the author and publisher do not assume and hereby disclaim any liability to any party for any loss, damage, or disruption caused by errors or omissions, whether such errors or omissions result from negligence, accident, or any other cause.

Adherence to all applicable laws and regulations, including international, federal, state and local governing professional licensing, business practices, advertising, and all other aspects of doing business in the US, Canada or any other jurisdiction is the sole responsibility of the reader and consumer.

Neither the author nor the publisher assumes any responsibility or liability whatsoever on behalf of the consumer or reader of this material. Any perceived slight of any individual or organization is purely unintentional.

The resources in this book are provided for informational purposes only and should not be used to replace the specialized training and professional judgment of a health care or mental health care professional.

Neither the author nor the publisher can be held responsible for the use of the information provided within this book. Please always consult a trained professional before making any decision regarding treatment of yourself or others.

GENERAL PAPERBACK ISBN: 9780578288055

Interior design by Matthew Wayne Selznick / MWS Media

ROLLING THROUGH
AFGHANISTAN

NICHOLAS BERTUCCI

Table of Contents

Dedication	*i*
Preface	*iii*
Chapter 1: The Influence	*1*
Chapter 2: The Journey Through Navy Boot Camp	*7*
Chapter 3: Navy Enlisted Medical Schooling	*11*
Chapter 4: Welcome to the Shit Show, Doc	*15*
Chapter 5: Iraq	*19*
Chapter 6: Welcome to Afghanistan	*25*
Chapter 7: Operations Begin	*29*
Chapter 8: Pressing On Through Hell	*45*
Chapter 9: Kinetic Encounters	*59*
Chapter 10: Wake Me Up When This Nightmare Ends	*83*
Chapter 11: A Month That Would Wake Anyone Up	*107*
Chapter 12: Mission Accomplished	*135*
Chapter 13: The Aftershock	*153*
Chapter 14: A New Life Mission	*161*
Chapter 15: Hell Comes Knocking	*163*
Chapter 16: Heroes' Hands In Healing Time	*167*
Chapter 17: The Final Message	*171*
Photo Gallery	*175*
Veterans Past	*199*
Veterans' Resources	*201*
Special Thanks	*203*
About the Author	*205*

Dedication

This book is dedicated to all my family, those who served in the Armed Forces, and those who support our Nation's Veterans. Without you, your love, support, guidance, and compassion, this all would not be possible. Thank you for giving me a reason to live and carry on.

Preface

This is my story to you for all those who have ever wondered what it is like to be a part of the United States Armed Forces and fight on the front lines in war. Many have their own version of their story to tell, and others may have trouble remembering things that have happened on deployment as they try to block out horrible memories that have scarred them in different ways. I kept a journal throughout my deployment in Afghanistan, and it became very therapeutic throughout the duration of my tour. Reading through the handwritten pages I carried with me in order to bring it to you in book form has been a difficult battle. I've suffered countless night terrors, shed a good number of tears, and experienced flashbacks. Still, it's been worth it to be able to pass on to others what it's like and for those who have gone through it. My hope is my experience will help others navigate the before, during, and after. I will not only explain how life was every day operating in Afghanistan as a United States Navy Fleet Marine Force Corpsman, but also life leading up to it, life after that, and the underlying long-term effects that are still carried to this day. This true story will take you through each emotion, battle, and physical feeling that have I've experienced day in and day out. It also explains the pains that are encountered as an after-effect and how it impacts not only the servicemember who went through the experiences, but also loved ones and acquaintances who encounter the scarred servicemember. This book explains the importance of recognizing the problem and what can be done by all to ease the pains from war.

Chapter 1: The Influence

My story begins where most would start considering the career option of whether to join the military. I was a student in high school in the town where I was born, Middletown, NY. My mother had signed me up to be a part of the Naval Junior Reserve Officer Training (NJROTC) Program. The best parental move she ever made for me as a teenager. I was still discovering myself at that age and probably would have had trouble in such a large and diverse school. I was mentored by amazing Naval Science Instructors that were paid by the school. Master Sergeant Thomas Willard; U.S. Marine Corps Retired, Petty Officer Meyers; U.S. Navy Retired, Commander Monty; U.S. Navy Retired, and Commander Andrew Wall; U.S. Navy Retired. These individuals played a large part in shaping and molding me toward a brighter future, whether I went on to join the military or not. I learned many notable traits; leadership, personnel and time management, communication, attention to detail, patience, risk management, and perseverance. I established myself within this program with many other great students who all positively directed me toward success. I had become best friends with many people who desired to join the military after completing high school. My best friend, James Allgor, and I wanted to follow Master Sergeant Willard and join the USMC infantry. After graduating, James went off to boot camp. I was denied at the recruiting station at the time because I had been on meds to help me focus better on my studies. It was devastating. I was stuck in my hometown for at least another year. Since I was set on joining the military, I had not fully prepared to become a college student. As a child, I wanted to be a marine biologist, and study and interact with sharks. But in community college, I took electives that struck my interest more. Plus, my biology professor at the time was terrible, in my opinion, and it wasn't easy to learn. The electives I took consisted of Basic Life Saving, and First Aid and Safety.

A year passed, and I was still living at home with Mom, Dad, and my two brothers Danny and Ronnie. I wasn't miserable; I was just unhappy because I knew that there was more to life than staying in the same town that I had originated from. I was working part-time at Subway, barely making ends meet. College was uninteresting to me, and I needed something more in life. I needed to escape and explore the world somehow. James had already come home from leave a few times telling me about his new military adventures, and I found myself living vicariously through him. On one fateful night, I came home from a rough day at school and work and found my father sitting in his room going through a big box of old photos. I greeted him as I was walking down the hall to my room; he called out to me. "Nicky! Come here, Son; I want to show you something." I stood next to him at the bed where he had laid out a bunch of photographs, some were in black and white, and some were an old, faded color. They were photos of his time serving in the Navy. He told me many stories that were all amazing; some were sad, some scary, but most were hilarious. He said something very important to me that night; it was a steppingstone that would forever change my life. It was the final push I needed to fly from the nest and do something meaningful with acceleration. He said to me, "Nicky, these were some of the best years of my life. If I could go back and do it all over again, I would in go a heartbeat, my son." It was at that point I knew what I needed to do. It all made sense to me. Many of my family members, such as my father; a Navy Sailor Petty Officer USN Vietnam War Veteran, Grandfather;

Navy Sailor Petty Officer USN World War II Veteran, Great Grandfather; United States Army, World War I Veteran, Uncle Dennis; Petty Officer USN Vietnam Era, Uncle William H. Keener; Navy Sailor WWII Veteran, Uncle Buddy; USMC WWII Veteran. It was my time to hoist my sails and carry on the honorable tradition.

The next day was an even more intense signal. I had to work at Subway again in the food court of the mall. It was always hectic there, and I hated it. In the middle of all the craziness during the lunch rush, a Navy Recruiter approached to order. It was a sign. He was wearing his summer whites, and I knew he was a recruiter because I recognized the distinguishing badge they wore for having that job. Before he could open his mouth and tell me what he wanted to order, he received an unexpected surprise from me. I assume he was expecting me to ask him what kind of sandwich he wanted. Instead, I looked him dead in the eye and said, "Give me your business card, sign me up, I'm joining." With wide eyes and a stunned look on his face, he replied, "Are you serious?" I replied, "Yes. I cannot be surer of what I just said to you. Get me out of here." It must have been the most effortless recruitment he had ever encountered.

I went to the recruiting station and went through the process of signing paperwork and taking the entry exam. The physical exam came afterward, and they were ready to ship me off once I had picked a job that I would be smart enough for in accordance with my entry exam score. It was time to break the news to everyone that I was getting out of town and what I had in store next for myself.

I broke the news to both of my brothers, Danny and Ronnie, first. They seemed indifferent about the idea and didn't really have a noticeable reaction. Danny did mention that he would miss me. I then told my mother. I remember it like it was yesterday. She was standing outside of the house on the front stoop. I had just arrived home from the recruiting station. Earlier that day, I had told her that I was going to stop going to Community College. She seemed upset about the decision that I had made. Once I got home, she was there outside and making small talk. She mentioned, "So, did you make any other big decisions today?" I replied, "Yes, I'm Joining the Navy." Tears began to form in her eyes as she trembled and asked me to repeat that. With a smile, I replied, "Mom, I'm leaving. I've joined the Navy, and I will be shipping off to boot camp soon." She cried some more, and I gave her a hug and told her that everything would be alright. She began asking questions like, "What about college, and your future to be close to home and be a marine biologist and interact with wildlife!?" I said, "Mom, I promise you that I will finish my schooling."

My mother was the sweetest and very sympathetic, loving person. She was always a great mom and loved her children endlessly. She was highly intelligent and had powered through her scholastic studies and achieved her master's degree in school psychology. I looked up to her a great deal when it came to achieving a college education. But the time for me to be a Sailor was now, and that would have to wait. The game-changer was when my best friend James found out. He contacted me as soon as he could. I've always known James to be a brilliant and strategic thinker. He always had a plan and knew his way around tricky situations. He called me one day in time before I had the opportunity to choose a career within the Navy. He said to me, "Bro, make sure you tell them you want to be a Corpsman." I said, "What the hell is a Corpsman?" He replied, "Trust me. Oh, and make sure you get it written in your contract that you want to be an FMF Corpsman." I trusted James and did as he told me. I researched and found out that I was signing up to go be the bad ass I always wanted to be. I was going to be with the Marines one way or another. This was the way. James had spoken, and I heeded his guidance. I became obsessed with the idea that I would finally be a part of something and have the best of both worlds within my military career. I finished out the semester, and it was time for me to ship out to boot camp. I said my goodbyes to everyone as the recruiter drove me to the airport. I was nervous. I was on my way to becoming a Fleet Marine Force Corpsman in the United States Navy.

Chapter 2: The Journey Through Navy Boot Camp

I was excited and nervous as I boarded the plane to Illinois. I felt as if I had left a world behind and there was no turning back. I arrived at the terminal where there were instructional signs directing new recruits to pool up in a location. An instructor greeted us and had us get on a school bus. As we traveled, everyone was silent. The bus was full of young adults like me that were about to embark on something that they couldn't imagine. I was excited and couldn't wait to get started. Once we arrived in Great Lakes Recruit Training Base, the busses parked, and another instructor boarded the bus. I was hoping he was going to start screaming and shouting at us. To my disappointment, he looked at everyone and calmly said "Get off the bus." We unloaded and stood in an orderly formation. They gathered us into several processing rooms where we went through the stages of sending out personal belongings and all clothes home in a cardboard box. We were all assigned out Navy gear and then to our divisions for training. I had researched what Marine Corps boot camp was like and what it was like to go through training as a Corpsman. The next two months were a joke to me. I was bored and already knew most of what they were trying to teach everyone as a group because of NJROTC. I knew the drill movements, commands, terminology, etc. I excelled and stood out because I became bored again. I felt as if I was becoming physically weak because I was going at the pace of those who weren't prepared for physical strain.

I remember asking my father what kind of person he was when he was in boot camp. He had told me that he was a troublemaker. I decided to have some kind of fun with this two-month long adventure at least to help pass the time. I decided to start messing with the instructors on a daily basis. There would be times where it was serious mode. We would all stand there at the position of attention, lined up with our bunks as the instructors barked orders about something and all of a sudden out of the blue, I would start dancing and smiling at the instructors. They lost their minds. My fellow recruits couldn't believe what I was doing. They beat all of us. When I say beat, I mean they made us work out. They started screaming at me and testing my strength and Naval knowledge. I was instructed to work out. They had me doing all sorts of different calisthenics thinking it would cause me pain and regret. I enjoyed it. I kept going until they got tired of looking at me. The other recruits could not keep up. They were exhausted and the instructors were tired of screaming. I was having a blast.

The only other recruits that could keep up with this pace of beatings were the ones that had signed up to become Navy SEALS - Sea, Air, Land. Those recruits thought it was hilarious and loved working out also, so they kept encouraging me to mess with the recruiters. The other recruits began complaining about what I was doing. I decided to slow it down because I didn't really enjoy seeing the others suffer for my enjoyment. When other recruits would mess up and be forced to work out, I would begin to take the blame for them so I could work out. The recruits began looking up to me with gratification thanking me for saving them. The instructors wanted to make me the leader of the recruit division because of this. I turned down the offer for the sake of allowing someone else gain leadership experience. The instructors began to catch on that I was taking the blame for the sake of excitement and wanting to work out. There was even one time where they had to pull me into their office and talk to me like a normal human being asking what my deal was. They were considering processing me out of the Navy for some sort of mental disorder. I could tell I had worn them out and they became very tired of figuring out how to gain the upper hand and break me. I saw this as a threat to my career and I had explained to them that I was bored and needed something more. I was ready to graduate and move on to the next phase. I was also envisioning becoming a Naval Officer already. The Chief in charge of the division decided to enter me into the dive motivator program.

This was a program within Naval bootcamp that would help condition those going into the special forces units; SEALS, SWCC (Surface Water Combat Crewmen), SCUBA (Self Contained Underwater Breathing Apparatus), and EOD (Explosive Ordinance Disposal) teams. I enjoyed this very much. I would work out for several hours in and out of the water with others who had contracts to become members of the special forces teams. Boot camp went by much faster after that. Before I knew it, it was almost time to graduate. When that time came, the dive motivator program instructors approached me with the intent to get me to relinquish my FMF Corpsman contract and sign up to become a special forces member. I was not for it. My desire was to be a Corpsman and be with the Marines. Plus, if I wanted to change my rate (job) in the Navy later in my career, I would. I turned their offer down. Once graduation hit, Mom and Dad came out to witness the ceremony. Dad mentioned that not much had changed on the training base since he was there in 1963. Him and Mom were very proud and supportive. It was a short weekend of liberty that went by and next was time for the next phase of training.

Chapter 3: Navy Enlisted Medical Schooling

Right across the street was Corpsman "A" school. It was a base dedicated to other rating schools as well, and the side the Corps School was on was very old. The buildings and barracks were outdated and small. It was like living in a very strict college dorm. Here in this school, there were seasoned Corpsmen that were granted the assignment to be instructors as a special billet within their careers. Classes consisted of 20-30 students, and every day was a classroom setting. Here they crammed about four years of medical knowledge into our minds within a three-month period. Each module had an exam at the end. There were about a dozen tests. If a student failed three tests, they would get booted from Corps School and become needs of the Navy. This meant that the student would not have received the rating they set out to have when signing up. The pressure was on. I was stressed out on a whole new level. I didn't think I was going to make it through because of how much we had to learn in such a short amount of time. Once it was finally over and I graduated, it was time to move on to the next school.

I quickly began to learn that within the military, friends come and go faster than usual. You go through struggles together and become close, but before you know it, it's time to move on, and you may never see that friend again due to their different military career paths. Some of the friends I have made I was able to look up on social media or have reached out to me to catch up fifteen years later. Some of them are still friends of mine, and it is fun to learn how they turned out. Some of these buddies of mine are doing very well and are still in the Navy. Others are no longer in, like me, but are still doing very well in other ways. It was difficult to adjust to, constantly saying goodbye to friends frequently, but there was nothing any of us could do about it. I learned that it was important to give people a great impression and never burn a bridge because you never know if they'll be your boss one day.

Before we all graduated Corpsman A School, each of us was given the option to pick orders to which coast of the country we wanted to go for the next phase of our training. I picked the east coast, and of course, I was sent to the west. I finally got a chance to take some leave and visit home before I was flown to Field Medical Training School. It was nice to see everyone again. Not much had changed back home. I was content with my decision to join the Navy. Some people were still doing the same things and living in the same areas. Only half a year had passed, but I felt as if I had already been through and accomplished so much since leaving.

After leaving, I boarded a plane and flew out to Camp Pendleton, California. It was exciting. I had never been so far away from home in my life. It was a whole new world to me. The base was extremely large, and we all got classed up after waiting for several weeks. In the interim, they had us doing yard work around the barracks. Each part of the base had a different area labeled by number. Field Medical School was exciting and intense. It was there where instructors trained us to be attached to and operate with Marine units. I wanted to make sure I got assigned to an infantry unit, so I made sure I gave it my all here physically. There were fewer restrictions than in A school. This was known as a "C" school. It was a specialty school within a rating. I was thrilled to finally earn my green and tan digital pattern camouflage uniform. We all looked like Marines as we stood there in formation and operated in the field. We learned the basics of combat medicine and gained basic knowledge of the Marine Corps, and discovered all of the basic gear that was used to save lives on the battlefield. After a couple of months of that, it was time for all of us to move on. Many went to their duty stations with the Marines or Battalion Aid Stations. Some even went home to serve reserve duty. I did exceptionally well within this school, so I was selected to go to a specialty school. It was a several-week-long course called Combat Trauma Management.

The training was way more in-depth and intense. It reminded me of a mixture of what I had heard about Marine Corps boot camp and Corps School with a field medic twist. A lot of the operations we did consisted of field maneuvers and swift combat casualty life sustainability. Sleep deprivation was a necessity here as well. They screamed at us a lot and had loudspeakers throwing out sounds of gunfire, explosions, screams from casualties, and chants from the enemy. They threw dirt on us as we crawled through mud under barbed wire and shot at us with airsoft guns. It was so much fun learning from this place. They did everything they could to stress us out and put pressure on us while trying to sustain life on the battlefield. It was an amazing simulation course that I would go back and enjoy doing again. They had realistic mannequins that bleed out and move, and the instructors screamed at us and nudged us while we worked on stopping the bleeding and started the breathing on our patients. There was this room called the blood room where they had ridiculous lights and techno music playing. The instructors even did crazy dances around the operation stations trying to distract us while we worked. Several weeks of this training were completed. It was finally time for me to be assigned to my Marine unit. I was nervous and excited to finally meet the Marines I would go into battle alongside.

Chapter 4: Welcome to the Shit Show, Doc

After our Combat Trauma Management Course was complete, we all gathered for a ceremony and were pooled into a briefing room afterward. Here we were assigned orders to our units. The instructor called out each of our names and told us which units we were going to be attached to. One by one, the names were called out. I was getting nervous again. Finally, my name was called. "Bertucci, you are getting attached with 1st Marine Division, Light Armored Recon."

I was ecstatic. Curious but also excited. "Oh, and they are deploying to Iraq in a couple of months, be sure to call your loved ones and be prepared." My heart dropped. This was the real deal. I was now pondering how I was going to die in battle. I had so many questions. We all did. The instructors caught on to our excitement and answered as many questions as they could from us. When it was time for a break, the first person I wanted to call and talk to was Dad. If I had called Mom, she would have fallen over in her chair, and I wasn't ready to deal with that because I was also trying to process it. When I broke the news to my father, I could tell he was nervous for me. He was my hero right then and there. He calmly said, "You will be fine, Nicky, you are strong, and you are great. Listen to your Marines and learn all you can. Keep your head low, Sailor; you will come home safe." With tears of joy in my eyes and the feelings of pride, I thanked him and told him he was right.

The next day, I was transported to the area where my unit lived and operated. I was greeted at the Battalion Aid Station (BAS) by my Leading Petty Officer, Doc Perez. He was a very large man who towered several feet over me. I was pleased to meet him, and he was pleasant. I looked up to him and said that I was honored to finally be here. He replied, "You have a very long way still to go; welcome aboard." We got to know each other, and he gave me a tour of the entire area. I stayed in a temporary barracks space to sleep for the night before I was able to meet the rest of the Corpsman crew and be assigned to my Company within the Battalion. Not to my surprise, the Sailors of the unit were seasoned and had already gone on one or more deployments in the past— all except for two other young Sailors fresh from training, CJ Calderon and Chris Daniele. We were the three new amigos who quickly became friends as we were assigned to the same company - Bravo Company. Doc Perez, our senior Line Corpsman, had already deployed once to Iraq and was placed in charge of us. We followed him around like little ducklings. This day, we were assigned to our platoons. Some of the higher-up Marines did their research and tried to find out who the new Corpsmen were so they could fight over who gets who. After getting the tour around the area where the company lived, I was instructed to move into my barracks room. It was a four-story condo-style building that overlooked the Pacific Ocean in the distance. I was assigned to the third deck of the building and had a Marine roommate. As I approached the duty hut of the company and entered the room on the first floor, I was greeted by the Marine on Duty. It was exciting to finally meet one of the Marines that I would be taking care of. He greeted me, saying, "What's up, Doc? Welcome to Bravo Company." I responded "*Semper Fidelis* Marine! Happy to be here!" I was happy to be there and I was very proud. I had finally made it to the life I had envisioned for so long.

    The Marines called Corpsman "Doc" for short as a sign of respect and because it was easier than saying Corpsman, and it also meant Doctor of Marines. *Semper Fidelis* is Latin for *Always Faithful,* and I could tell they liked hearing me say it. I was young, wide-eyed, and motivated. I was eventually assigned to Blue platoon, where I met the group of 20 Marines that I would be working directly alongside. It wasn't until the next morning's formation when I was properly introduced.

The next morning, the Marines formed up in front of the barracks. This was where I was able to witness the entire Bravo Company. Our Staff Sergeant was standing at the head of the platoon facing us as we formed a square. I was standing in the center of the Marines, where the plan of the day was ordered. At the end of his speech, Staff Sergeant looked at me and smiled. He then said to the platoon, "We have a new Corpsman with us, men. Let's give him the Blue Platoon warm welcome." As he finished saying those words, the Marines turned inward, all facing me with shit-eating grins on their faces. I stood still with a look on my face as if I were about to enter a prison yard brawl. My brow lowered, and my eyes looked around at them. As I began to smirk and before I could speak, the Marines all took punches and kicks at me after a couple pushed me to the ground in the middle of the pile. I yelled out, "Aww fuck you, assholes!" as we all laughed and kicked up dust. It was my welcome to the crew initiation beating. I was happy to be there, as if I was with a pack of brothers who automatically loved me. The beating was short, but when it ended, the Marines would walk away a few at a time, saying *Welcome to the team, Doc!*

The Marines did their routine, and I became oriented on their activities while sharpening my skills as a medical professional at the Battalion Aid Station during sick call hours. I would bounce back and forth between working with the Marines and working at the aid station learning all that I could. It was sometimes a battle between worlds because some days it was difficult to balance being in two places at the same time. When it came down to it, when my Navy Chief or Leading Petty Officer gave me an order, I disregarded the orders from the Marines. After all, the Navy higher-ups were the ones that wrote my eval for promotion. Things went on like this for several months until it was time to do our pre-deployment workup. Some units called this month-long field trip Mojave Viper. It was a training simulation where we would convoy as a battalion to the USMC base 29 Palms. It was located in the Mojave Desert. We went through several range operations and close combat simulations where we were all tested and scored on our abilities to carry out combat operations. It was pretty shitty living conditions, but the training was good. All operational units were required to do this pre-deployment work-up before deploying. There was a lot of good training that took place here, and a lot of blood and sweat was shed. Some bones were even broken from time to time. After Mojave Viper training, several more live-fire range exercises were conducted. A few weeks before deployment, the unit was granted pre-deployment leave. We would all visit our homes or wherever we came from to spend with family before embarking on a seven-month deployment to Iraq.

Chapter 5: Iraq

It was nice seeing family and friends again. I was nervous about deploying for the first time to a war zone. I was a boot, so I knew I was going to get messed with a lot. Myself, the other boot Corpsmen, and Marines shaved our heads clean for the tradition of letting everyone know we were boots on our first deployment. Some didn't want to, but often times in the USMC, you don't get a choice. Once we arrived in Iraq, it was a similar environment to that of Mojave Viper. Hot, dry, and full of sand. We would ride around on the Light Armored Vehicles conducting reconnaissance of different areas and villages in the vast desert and stopping vehicles and questioning the locals in the Al Anbar province. We also helped train the Iraqi police. The Iraq deployment wasn't as kinetic as I had imagined. There was a lot of sitting around and occupying territories. We spent a lot of time at the Al Asaad Air Base, where they had basic amenities such as entertainment centers, restaurants, shopping centers, and good chow halls. This place was established really well for quite some time before we had reached it. Much of the action was taking or had already taken place in the major cities of this country, and we were far from any of it. A big pressure for the junior Corpsmen was getting our warfare device. It was known as the Fleet Marine Force (FMF) pin. It signified our service with the Marines and that we had mastered the knowledge of this inch-thick book given to us to study. There was a lot of reading, studying, and quizzing in preparation for the board. We had a lot of time on our hands, so we were expected to read the book verbatim on the board.

Iraq was very boring. Especially after the Status of Forces Agreement came about a quarter way through our deployment. This agreement was the end of the war and the start of U.S. forces pulling out. We sat around holding a position in an abandoned castle in the middle of the desert for the rest of the deployment. We would do short perimeter patrols every couple of days just to check the surrounding area. This castle was cool. It consisted of two large structures made of stone. We called the position H1 because it had a helicopter landing pad in front of it. Behind it was a factory that no one ventured into. Near the factory were small living huts for the workers, we assumed. A lot of farmers and sheep herders would stop there for shelter throughout their travels. There was sheep shit everywhere. The flies throughout the summer were atrocious. Millions of flies constantly. It was terrible. We would place fly strips out, and within a couple of hours, they would be completely black, covered in flies. If you were to get close to it, they would sound like chain saws buzzing at full throttle. Behind the workers' huts there was a gravesite. The Iraqi people wouldn't bury their dead. Instead, they would lay the bodies on the ground and pile dirt on them and decorate the piles with rocks. These were called cairns. A tank battalion got attached to us one day. Out of nowhere, we could hear thunder off to the distance in the vast desert beyond our position. Standing outside of our living spaces, we saw the cloud of dust being kicked up as the rumbles got louder, and then we could finally see the tanks rolling in from behind the factory. I had no idea why we had a tank unit attached to us at this point, but it was cool to see tanks in person and learn about them firsthand. As the tanks got closer, they looked as if they were headed straight for the gravesite. I assume they had no idea what they were about to do. As we watched, the tanks ran over all of the bodies within the gravesite. It was shocking to see. Once attached to us, they conducted the same important business that we were up to throughout our time here, sitting around doing nothing.

The weather began to get cold as the winter approached. Eventually, it became freezing, but we never got snow because of the dry climate. Soon after that, it was time for Calderon, Daniele, and I to test out for the FMF pin. It was an easy board because when we finally got picked to actually do the board, we could read the FMF book front to back. It was nice to never have to pick that book up again. We were all sick of it. The ceremony was very nice, and I felt confident as I was pinned this special piece of medal on my chest by my mentors. I don't really have anything positive to say about my Iraq deployment other than that. Nothing else was significant that I could remember, and I would have a very hard time keeping a journal of events. In summation, Iraq sucked ass. I learned how to stay sane despite boredom. The seven months seemed to go on forever. It was like a new form of hell that I would never want to revisit. SSGT McCormick made life seem a whole lot more pleasant out there because he was very good to us. He connected to us on our level and led his Marines with compassion.

My ability to function with the Marines and operate with them had been proven. It was time to come home. I came home feeling accomplished in the sense that I was no longer a boot. It was strange being home, but none of us had any noticeable issues yet; there were no casualties or traumatic events that took place, except when we had to kill an Iraqi driving at us head-on and wouldn't stop his vehicle. There was no time to react because of how quickly it happened, so the escalation of force had to be ruled out for the sake of the safety of the convoy. There was a huge investigation that took place by NCIS because it wasn't reported up by a sergeant who made the call to take this Iraqi's life. My parents were concerned. I let them know I was fine and there was nothing to worry about. They wanted to make sure I didn't have PTSD. I was just a little disgruntled from all the bullshit I had to put up with for seven agonizingly boring months. There would be some nights where I would wake up looking for my rifle, thinking that I was still over there. Once I realized I wasn't over there anymore, I would calm myself and try to fall back asleep. Being in public was strange as well. I remember going grocery shopping and feeling as if something was very strange. I would always be on the alert because of how long we had to operate like that. It became second nature to carry about as if the enemy were about to ambush at any given moment. Once I returned back to base from post-deployment leave, the company started changing. Marines I worked with were getting out of the Corps, new ones were arriving, and my fellow Corpsmen moved to different companies. I was voted to stay with Bravo. They nominated me to be the Senior Line Corpsman and take Doc Perez's place while he moved on to a new command elsewhere. I was honored that I had been chosen. Calderon and I were also chosen to go to advanced training. We stood out as motivators, and the Marines wanted to make sure we were equipped with mountain medical knowledge in preparation for our Afghanistan deployment that we didn't know about yet.

Some months passed, and the company was whole again, with new personnel. One fateful day, the new company commander called us to gather. He gave his briefing about upcoming events. He also added at the end that we were slotted to deploy to Afghanistan. We were all silent. It was very kinetic over there, and we all knew it. We knew that some of us might not come home from this one. In preparation for this deployment, our gears shifted toward mountain training due to the harsh terrain over there. I was sent to specialized training. Mountain Warfare Medicine was a course designed to prepare units for Afghanistan. It was a very small school located in Bridgeport, CA. I was one of the few that were selected to embark on this training. When we arrived, there were mountains as tall as the eye could see. It was a beautiful place that reminded me of home. We were there with the British

Royal Marine Commandos, and they were bunking on the other side of the building we were staying in. We quickly made friends with them and exchanged pieces of gear. Great group of men. They had awesome senses of humor and would even sunbathe outside of the barracks on a cloudy day because they mentioned that Europe was always rainy. Throughout the duration of the training, we learned how to maneuver casualties from rugged mountaintop terrain, across freezing streams with rope bridges that we made, repel upside down from tall cliffs fully loaded with gear on our backs and rifles in our arms, and even submerged in a freezing river to get first-hand experience on hypothermia treatment and rewarming methods. I loved every bit of this training and would love to do it all again. They even offered a cold-weather course during the winter that looked very difficult, judging from the pictures they had on the walls of the schoolhouse.

Once training was complete, we traveled back to our units. The Marines were prepping for a new round of Mojave Viper training, and the BAS was preparing shots and all records for deployment. I broke the news to my family when I took pre-deployment leave. They were all scared for me. I was nervous and excited but was worried that kidney stones I'd developed from the MREs in Iraq were going to disqualify me. I decided to suck it up and press on.

We convoyed to Mojave Viper again for more training with the incorporation of mountain warfare training, more live-fire exercises, and casualty maneuvers. The chow halls there sucked, and it was hot as hell. Good training. I was doing all that I could to train my junior Corpsmen on everything I learned. They were a great group of young men who got along well with their Marines. Once another month of Viper training was complete, it was time for pre-deployment leave again. I enjoyed spending time at home with family and friends, and before I knew it, it was time to come back to base. I flew back to California from New York with my head held high and the idea that I may never return home again. It was difficult to deal with, but I had a mission to focus on and preparations that had to be made.

Chapter 6: Welcome to Afghanistan

The Marines and I loaded up on a huge airplane and flew out to Kuwait. Just like Iraq, it was the holding point for all units flying in and out of the Middle East. It was early spring there, and it was 120 degrees in the shade. We laid around in a huge building filled with green mesh cots trying to acclimate to the weather. The most exciting thing going on for those few weeks was eating at the chow halls and watching the new boot Marines get screamed at. We finally got slotted to load on another plane. An interesting fact about this flight was that our pilot's name was Captain Morgan. All the Marines cheered once he stated his rank and name on the intercom. It was hilarious. What was even more hilarious was that none of us would get to drink any Captain Morgan for the next seven months. Once it was time, we pressed on to arrive in Afghanistan at a huge base called Camp Leatherneck. This base had it all as it was well established away from any combat areas.

We stayed here for a couple of days as an entire unit waiting for the next plane to take us to our destination. While we waited, we took several classes on IED recognition and different operability tactics. One day, we were sitting on the bleachers under a shaded tent as an entire company of Marines and Corpsmen. We were awaiting the instructors to show up for what seemed like an hour or so. As we waited, one Marine decided it would be a good idea to entertain the company by getting up and showing off his tattoos to everyone. This became a show and tell event as the rest of the Marines, one by one, presented themselves to show off tattoos. Some Marines had to take off their shirts, and others had to roll up their pant legs. As the Marines would show off their tattoos, the ones in the bleachers would give out cheers or jeers. The showstopper was this one special Marine named Lance Corporal Seth Carlson. He walked up with pride, boasting that he had the best one of all. As he began to undo his belt from his pants, he swiftly turned around and bent over. Not only did this young Marine moon the entire company, including the higher-ups standing amongst the crowd, but he also spread his ass cheeks exposing his sphincter as the Marines all screamed at him, and while others laughed so hard, they fell from their seats. He was a hero in my book for this. It was the perfect opportunity and display of jackassery that I still laugh about today ten years later.

The plane finally landed. Once we all made our way to the doors of the plane, I could hear several people complain about the air quality and climate. Imagine heating an oven up to 150 degrees, then opening the oven door and sticking your face in. I could tell that it was going to suck trying to acclimate to this area.

Once we all unloaded from the bird and were rounded up, we were shuttled to tents, where we stayed for a few days until smaller aircraft transported us to our main forward operating base (FOB). This new base was called Camp Payne. It was freshly established and was approximately a 100-acre square surrounded by sand-filled Hesco barriers and watch towers with Marines on guard. The chow hall was a large green pop-up tent, and everything that required electricity was run by generators. We were herded like sheep carrying all of our gear to smaller green tents that had no insulation, no flooring, and no air conditioning. It was hotter in the tents than it was outside. We were all tired from our trip and drenched in sweat. The showers here were made from mobile trailers that would pump hot water only from the tanks sitting in the sun outside next to them. I was happy that there were at least those things here already. Little did we know that we would seldom see this base throughout the remainder of the deployment.

It was the end of May in 2010, and we were all sitting around in the tents for several weeks, acclimating to the climate. It was already miserable. Some of the Marines were doing this thing called left seat right seat with the Light Armored Recon unit that was already here. Our vehicles were staged down at the ramp at one end of the base, and there were several wooden hut-style buildings erected for the command higher-ups to have their meetings and go over missions, as well as one for the BAS. I checked into the BAS and found Sailors from my unit that had been there already for several weeks. They gave me the rundown on how things operated out here, and I would bounce back and forth between the Marines on the ramp and conducting sick call during those hours for the sick and injured Marines at the BAS for several days. I was also getting loaded up on medical supplies and medications in preparation for operations outside of the wire. Operations wouldn't begin for us for a couple more weeks as we prepared the gear on our vehicles that had been transported out here. We also had to do a proper turnover with the Marine unit that we were relieving.

Once this was all complete, operations had begun. We would venture short distances outside of the wire (base: Camp Payne) on patrols in order to orient ourselves to the area. Once we left base on patrol missions, we had to go down a ramp and cross over the Helmand River. This river was muddy and small, but it was deep and large enough to require vehicle transport by boat. As we approached the riverbank, flat ramp boats were there waiting for us to drive on. To everyone's surprise, these boats were operated by the Army. We would don tarnished sun-dried life jackets before crossing. Once the convoy was across, I could see the base fade in the trail of dust that we left behind. This was it. We were in Taliban territory, and it was game on. There was a small, more forward operating base down the road called South Station. It was much less established than Payne and consisted of large tents as living quarters. It was a much smaller square that consisted of about 5-6 acres in size and was surrounded by Hesco barriers and guard towers at every corner. As we drove past, we would all say, "Man, what a shitty situation it would be to have to live there for any period of time." There was never any shade unless it was man-made, and the dust here was so fine that it would get into every crack, crevice, and orifice of every single part of every single thing that it touched. It would always pile into the back of the scout hatches of the Light Armored Vehicles (LAVs) as we rolled on because of the backdraft. There was no escaping being caked with piles of dust that would stick to us in layers as we sweated profusely throughout the day. I decided, since this place was presumed to be a kinetic adventure of combat and excitement, that I would keep a journal log in the Write in the Rain note pads that I had on me. I was going to write every day about events so that I could share them with loved ones back home. I never had any intentions of creating a book such as this, but once the operations started, I kept track as we rolled through Afghanistan.

Chapter 7: Operations Begin

June 12 – The first operation was a late-night movement. Just a few minutes into our patrol, one of our 7-ton vehicles carrying supplies hit a sand berm and flipped over. There were two Marines that we had to pull out after the vehicle had rolled. Two casualties and leaking fuel. We called for the recovery vehicle to deploy and assist. We stood guard as a group of Marines worked to recover the vehicle. I did examinations on the Marines that we pulled from the flipped vehicle, and they were both okay with minor scrapes and bruises. We continued our patrol out in the vast desert, learning about the area and certain landmarks until it was time to set up a campsite for sleep.

June 13 – We started our day with very little sleep. We patrolled for most of the day only to have our first two encounters with the people living in the area. We spotted and stopped a truck with sticks in the back of it. It took forever because the sticks were piled ten feet high from the bed of the truck. We searched it with the suspicion that this Afghani may have been hiding something underneath. We found nothing significant and let him pass after helping him load all of his sticks back on the truck. Our other encounter consisted of a small truck that, no shit, had 26 sheep piled on the back and 4 people inside. These Afghanis really knew how to pack weight onto their trucks. Transportation was very scarce for the locals, so they must have made the most out of every trip. Again, we found nothing after searching the passengers and making them unload every sheep in order to conduct a proper search of the vehicle. We continued to patrol the desert passing small villages along the way until it was time for rest at another hastily made camp site.

June 14 – Today was a day of observation. The location where we set up camp last night was where we remained for the entire day. I saw my first dung beetle today. It was rolling a ball of camel shit through the sand, and I watched it with amazement. Fun with dung beetles. There were flies everywhere in massive quantities. It was extremely annoying. There were sheep farmers everywhere, which produced sheep shit and, therefore, the flies. Iraq was the only other place where I've encountered so many at one place. The rest of the day consisted of hanging out in the ring of vehicles parked in a circle facing outward, which we called a coil. It was a slow yet gradual start to our operations on this deployment, but it was good things weren't hot and heavy right off the bat.

June 15 – We snapped two men on a tractor passing by and found 26 kilos of raw opium. We detained the two individuals and took their information as we waited for other members of the command to reach our position to take them and the drugs back to base. I conducted physical examinations and took notes of scars, and identifying tattoos if they had any. They wreaked of body odor. After they were taken for transport, we sat around all day in the hot sun doing nothing. By noon, I was craving McDonald's for some strange reason. It would be another seven months before those and many other tastes became familiar again. As we were going about each other's business within the coil, one of the watch-standing Marines spots another tractor trying to evade us. We sprang into action and took in two more detainees. I conducted physical examinations on them, and members of Blue platoon found 4 lbs. each of heroin and ecstasy hidden inside the tractors.

June 16 – We drove around the desert some more in the morning. We then stopped to set up an ambush point several miles outside of this place called Camp Rhino. It was the first base established after the Afghan invasion. It was eerie, deserted, and most likely covered with IEDs. We set up an observation post on top of a hill that could overlook the area and waited there for the rest of the night, scanning with our night vision goggles.

June 17 – I learned a new card game called Egyptian War while we sat around waiting for our turn for the observation post. I played with some of the Marines until 15h00, when we mounted on the vehicles and left the outskirts of Camp Rhino to go resupply. I got a letter with pics from an old friend from back home and two postcards from another. One of the sergeants fell and suffered a head injury. I guess he was sitting down low in the vehicle when a computer fell from the top of the hatch into the hole, hitting his head and causing a minor laceration. I cleaned it out and closed it with a couple of butterfly Band-Aids.

June 18 – I got a chance to read the letters that were sent to me. I had to wait till the next day because we weren't allowed to use lights at night that would give away our position. One of the Marines chipped off half of his front tooth while cranking wrenches on the vehicles. I managed to go through an entire bag of sunflower seeds already. 11h00, we left to move to a new position, and while on the way, White Platoon reported 2 white pickup trucks, one still running with the passenger door open and no one inside. We proceeded to White's position and found one man walking away from the trucks. We captured him. We then proceeded to search for more through the sand dunes and detained 2 more men running from trucks in another direction. The trucks had 2,500 pounds of methamphetamines. Another truck with more meth and three more detainees were captured by White Platoon. We had a whole mess of work to do this day. The men transporting the paraphernalia must have spotted us before we could spot them and decided that they could escape on foot. We were far outside of the wire and called in for transport of the detainment. It was going to take a good while before the seven-ton vehicles reached our position, so we set up a coil there for the night.

June 19 – 0200 I was woken up for detainee watch. They smelled like a mixture of old people, body odor, and cow shit. We sat around all morning until Crazy Train (the seven-ton vehicles that would transport equipment to and from our operating positions) came to get the detainees and trucks with drugs. This was one of the hottest days out here so far. I got the opportunity to call home, but it was 2 AM their time. There were no answers, so I left a few messages to several folks letting them know I was alive and well.

June 20 – It was way too hot to sleep last night. I woke up feeling tired, dehydrated, and fatigued. We resupplied Red Platoon and then drove around the desert orienting ourselves to the new area. While traveling to a new campsite, we crossed paths with a man and his son driving a white truck. We dismounted to approach and question them with our interpreter working with us. They tried to run from us after seeing us on foot. We then remounted the vehicles, put the pedal to the floor, captured them, searched the vehicle, and found 200 kilos of opium. As we proceeded to detain the two, they claimed the Taliban made them drive the truck. These cases out here were true on many occasions. Members of the Taliban would terrorize farmers and locals, threatening to kill their entire families if they refused to conduct operations toward their cause. As we started unloading drugs from the white truck, Red Platoon reported two dark trucks in the distance that took off after realizing we were in the area. We split up and took three vehicles to help Red Platoon track them. While doing so, an Afghani fire ant crawled in my uniform and bit the shit out of my arm. I panicked and stripped off my flak and fire retardant over garment (FROG) top to see 5-6 bite marks with localized edema and erythema (swelling and redness) surrounding the bite sites. I took Benadryl and rubbed on the ointment. A couple of hours rolled by, no sign of any other trucks, and air support that we called in had no luck as well. We returned to the coil with the rest of HQ (headquarters) platoon. All the drugs we confiscated were laid out for inventory by the time we got back to the coil. The rest of the Marines were taking pictures, posing in front of over 200 bags of opium that were the size of regular pillows. Total estimated worth 250 million dollars. I proceed to do my Corpsman duties and conduct physical examinations. It turns out one was a 32-year-old man with asthma and one kidney. The other was a short 20-year-old. Both men were members of the Taliban. It was now 19h00, and we were awaiting Crazy Train to punch out to our position and collect all that we captured; drugs, a truck, and two detainees. I was feeling drowsy from the Benadryl I took earlier today, but it was still too hot to fall asleep. I hated waking up feeling dehydrated. Later that evening, I got a chance to call Dad for Father's Day. The night went on hot as hell, and the Marines and I did our duties throughout the night taking turns standing security and detainee watch.

June 21 – Early that morning, the seven-ton trucks arrived at our position, and we rid ourselves of our bounty. We then started driving around the desert, looking for more Taliban to take down. It was 10h00, and already hot as hell and flies everywhere being annoying pains in my ass. Red Platoon found an abandoned truck with 163 kilos of opium inside. The truck drivers fled in another vehicle and were nowhere to be found. HQ met up with Red to retrieve the truck. As we headed back to camp, we decided to abandon the drug-filled truck and set up an observation post behind some hills in the distance to lure more Taliban in, Trojan horse style. Loui O (my vehicle driver) and I decided to break out the fly swatter and get sweet revenge on those pesky flies after growing bored of mutilating the shit out of several dozen flies, ripping off wings, legs, and hocking loogies on them as they tried to crawl away. We chilled in the LAV for several hours. We then got swarmed with another wave of flies that were more annoying than ever. No doubt they were seeking revenge for their mutilated and tortured dead friends. As the sun began to set, I worked out and then went to bed.

June 22 – 02h00 Woke up to stand post for 2 hours. There was no suspicious activity to report during my time on post. HQ packed up camp and prepped for the next operating post. Some of the vehicle commanders were arguing back and forth on the intercoms about who has who's binoculars. I witnessed one of the sergeants completely lose his mind and chew a Marine a new asshole over misplaced binoculars. We then resupplied Red Platoon after traveling to their location several miles away. We still had the abandoned truck filled with drugs and set up another Trojan horse ambush with it in a new location. I was sweating my ass off while swatting flies for several hours. As we provided security for Red's resupply, Black-6, (another vehicle of Marines in our platoon) reported a truck heading east towards our position. The truck saw us, turned around and headed east at a high rate of speed. The chase was on! HQ flanked the truck's route, and we caught up, waiting to ambush. Meanwhile, Red Platoon was overwatching the Trojan drug truck. Loui O began vehicle maintenance while I proceeded to pick a pound of dust out of my nose and swat at more flies. 13h00, and it turns out that truck was just a tractor, so we said fuck it and decided to leave it be and coil up. I sat around playing cards with Loui O for a couple of hours. Another tractor rolled by, we snapped it, nothing significant to find. Snap is a term we would use within the Marine Corps when a team would stop a vehicle and search it for paraphernalia. Several more tractors rolled by that we snapped with just farming goods as 1st Sgt, Loui O, and I passed the time playing card games and cracking jokes. After the sun went down, gunny surprised the platoon by inviting everyone over to his pig (we would also call the LAVs pig) for some delicious tacos. I made my appearance saying, "I'm here to kiss the cook." Laughter broke out. The tacos were amazing. Soon after, it was time for sleep.

June 23 – 02h00 post again. P. Diddy (one of the young Marine vehicle drivers for Master Sergeant's vehicle) came over for card games for a while, and then I caught a gecko lizard. Let me tell you about P. Diddy, aka Powder Smash. This Marine was difficult to deal with. He used malingering, skating, self-misery, selfish manipulation, complaining, and frequently tried to weasel his way out of most things. How he even made it through boot camp simply boggles the mind. The drill instructors must have pushed him through the system to graduate because they couldn't fix him. Time went on sitting around. Eventually, I caught another lizard and made lizard earrings for my vehicle driver Loui O. It was a form of boredom entertainment where one would catch a lizard, piss it off and make it bite, and hang on to the ear lobe resembling a live earring. Some more time went on, and we packed up camp. 09h00 hot again already. Thank God, cloudy and overcast today, so not so hot anymore. We continue our mounted patrols while a sandstorm pounds on us. The dust in this place is so fine and gets into every crack and crevice of every single part of everything that exists. Fuck. I'm sure our lungs will be damaged beyond repair at the end of this tour or eventually develop into something severe as we grow old. We coiled up again after some time patrolling in the desert, visibility was down to about 3 feet, and the air quality turned red. Red air meant that since the visibility was so poor, casualty evacuation methods by air lift couldn't be conducted because the helicopters weren't able to fly, thus forcing ground units to temporarily halt operations. Soon it began to rain. Everyone is confused and excited about this odd, random weather. As fast as it started to rain, it died down within 5 minutes. HQ platoon was on the move again. We drove around some more, set up a couple of coils, then we finally got to the last one for the day. I helped Loui O change his planetary fluids (which are the fluids that keep the rims cool and lubricated on the LAV). Shortly thereafter, we discovered that we had some more time to kill, so he taught me how to drive the LAV. It was like driving construction machinery. The driver's "Hole" was about as wide as my hips and narrow with no airflow. To my right was the turbocharged motor that would scream in my right ear and radiate massive amounts of heat with the press of the gas pedal. It was interesting looking through the rectangular periscope over the top of a massive military vehicle. I always wanted to drive one. That was to be the highlight of the day as the sun dropped over the horizon. More time passed with nothing going on. Time for bed, but I was woken up in 30 minutes to check out what appeared to be an STD on a Marine's penis that was causing pain and bleeding. Come to find out that this particular Marine had mutilated his penis with a pair of pliers to make it look as if something was terribly wrong. I was under the impression that he was trying to figure out a way to escape deployment and go home. At the very least, he was attempting to be placed at a base where he did not have to do very much work and not hate life as much. Life was better on a forward operating base where there were certain amenities such as a commissary, showers, phones, computers, and internet. This place was called Camp Leatherneck. He could have had intercourse with something disgusting out here, but the chances were slim. I sent a BFT (Broadcast File Transfer) message up to the BAS (Battalion Aid Station) back at Camp Payne. The medical officer suggested we send him back to get checked out because there was nothing I could do but give him Advil and antibiotic ointment. I went back to sleep. I was woken up 30 minutes later to stand fire watch for an hour and a half. It's 0h10. I was tired and eventually got relieved from post and tried to fall back asleep.

June 24 – Fuck, I am tired. Logistics Crazy Train punches out to resupply the rest of the company. We began unloading the water and chow at 07h00, already sweaty as fuck. HQ

Black-6 vehicle finds and uncovers an IED while on patrol shortly after the resupply operation. They set up a perimeter around it and called the EOD (Explosive Ordinance Disposal) team. Shortly after we spotted 3 cars moving from the IED site, we intercepted them and found jugs, some electrical wire, and a solar-powered battery in the cars. We proceeded to tactically question each member of these cars and in order to uncover their stories about what they were doing out here. We suspect cars to be rigged with VBIEDs (Vehicle Born Improvised Explosive Devices) so carrying on with extreme caution was a necessity. After conducting a thorough search, there was nothing else significant to find in the vehicles. Without any damning contraband, we could not detain these individuals despite the fact that all signs pointed toward them hating and wanting to kill us. One man had a set of keys, and one of the keys had been filed down. We suspect he was going to try and hijack our Trojan truck that we had set up for a sting operation. (This truck was a vehicle we had found earlier with drugs in it that had been abandoned.) We decided to leave it as is and watch from a distance to see if the operator or transporter would return in the case that he saw us and fled in order to avoid capture. 09h00 took a nap and woke up at noon. I had an awesome dream about this girl I was talking to from back home in NY. Also, I had a dream about Dad running a marathon. My dreams were strange while on deployment. Sometimes they were good, and other times they were obscure and bad. [Little did I know then, but they would eventually worsen to the point where it would affect my everyday life after I came home.] Some time went on burning the day. We sat in a coil to resupply Bravo Company. We had some downtime and got the opportunity to pass the satellite phone around. I called Dad, my brother Ronnie, a friend, and Chad today. It was good to hear their voices, although it was only for a brief few moments. Missing home was becoming more intense and more frequent. I then hung out with the Black-5 vehicle crew.

No sleep yet, only short naps. 23h30, a C130 air drops chow, water, baby wipes, and unitized group ration (UGR) meals. My favorite UGR was the breakfast meal because it tasted the least like cardboard and dog shit. It was a fun sight to see. Come to think of it, anything out of the ordinary (ordinary being staring into the vast desert hills of nothingness) would have been fun at this point. I would get so excited about the simplest little things, such as seeing a dung beetle scurry across the sand. Once all of the boxes were dropped, there was a huge mess scattered everywhere from the parachutes and several busted boxes from the 800-foot fall, bottles of water were spread all over the desert. We started cutting the parachutes loose from the cargo boxes and piled them up to burn. We started by piling the boxes in the 7-ton supply transport vehicle until we got down to the individual scattered bottles one by one, throwing them in the 7-ton. At this point, everyone was insanely tired and moving like zombies but still somehow finding a way to have fun--03h00 rolls by. Everyone passes out on top of the massive garbage pile. We set up a fire watch schedule and catch some sleep for a couple of hours. We wake up at 06h00 and pour gas on the pile. LCPL Brown lights it up. This flame was huge. Smoke filled the air, and we soon left the coil and got some rest.

June 25 – The same place, still vainly trying to get rest, but it's too hot outside. I would lay on the highest point of the vehicle to try to catch the slightest breeze even while leaving myself completely exposed to the enemy. I simply gave very few fucks at this point. There were times where I felt that I would rather die than have to continue on living this hellish nightmare we called deployment. I then stood radio watch on C2 (communications vehicle) and learned how to work satellite maps and send secure Byzantine Fault Tolerant (BFT) messaging. I remember hearing an interesting conversation with the XO of the battalion. Not sure what it was about, but it was like eavesdropping on the bosses.

June 26 – I got mom's package with flavored drink mix packets. I got my 5th letter from that girl I had a dream about a few days ago and a letter from St. Joseph School 2nd grade class which really made my day. Earlier I caught a small scorpion crawling on the ground next to my flak jacket. I showed it to 1st Sgt. in the water bottle I had emptied out to catch it with. He said it was highly toxic. It was black with a fluorescent green underside, and the stinger was fluorescent green. A couple of minutes later, I felt something crawling up the right side of my neck. Thinking it was a fly, I swatted at it and still felt it crawling, so I pinched it off, looked at it noticing it was instead a tick. It was a brown-bodied, yellow-legged one. I was extremely surprised to learn that they even existed out here. Later that day, Loui O had one crawling on his shirt, and I caught it in another water bottle.

"The river just knows" by Rodney Adkins is a good song for me. [I guess I heard that song while sitting around doing nothing on one of the vehicles that day, and it calmed me and reminded me of home.] The day goes by slowly, trying to stay in the shade. I notice myself getting frustrated with this deployment; little things people are doing are starting to annoy me. Some examples are: Powder Smash deceiving me and taking advantage of my job as a Corpsman just so he can get out of the field for a night, SSGT Chantickle being an annoying asshole and trying to fuck with me by bossing me around just because he is a higher rank, the heat of the day, the dirt and crusty salt all over my body from constantly sweating and not being able to shower for weeks on end, not being able to communicate with people from back home when I want to. It has been about 2 weeks since I've talked to anyone. My mind is made up about re-enlisting- Not happening! Well, it is time for rest. I am going to try to rest my troubles until I get woken up at another ridiculous fucking hour in the middle of the night to stand watch. This time I will be in a flak jacket walking around like a fucking scout Marine even though I am not supposed to fucking do this shit. Everyone thinks it's funny when I tell them I am technically not supposed to do that shit because I am a noncombatant. What-the-fuck-ever-man. I received a direct order from my Chief not to stand gun watch. Thank God the day is over for a couple of hours. 30 minutes later, Gunny Manny runs up to me and tells me one of his mechanics has a piece of metal in his eye. Finally, something interesting going on. The Marine comes over to me, and I flush it out with warm water. I have a two-and-a-half-hour roving watch in two hours trying to determine if I should go back to sleep or not. I didn't go back to sleep and stood 3 hours of watch due to the watch schedule being so fucked up.

June 27 – Woke up after 3 hours of sleep for reveille. An hour goes by, and there I am standing post yet again. Another hour goes by before Loui O relieves me. I try to take a nap waiting for resupply Crazy Train to come in. Doc Mantis comes up to me presenting a patient with methicillin-resistant staphylococcus aureus (MRSA) in his nose. MRSA is a bacterial infection that is drug-resistant and can lead to death if not treated in a sufficient amount of time. It starts out as a staph infection that, in many cases, is derived from cellulitis, which is pretty much a gnarly super infected pimple. I made the decision to CASEVAC him back to base for a higher echelon of medical attention. It must have been 120 degrees today. I mean, it has been miserably hot all day. Everyone has been acting delirious from the heat and agitated from sitting around doing nothing all day. Later on, it was my turn for post on Black-6 at 17h00. I get off an hour later, awaiting our night operation at 20h00. The sun began setting, and it is starting to cool down, praise the Lord. I fall asleep for the whole ride. I wasn't sure where we were going. Most of the time, I was unaware of where we were going or what exactly was going on. I was just along for the ride with the purpose of keeping my Marines healthy.

June 28 – It is very windy out; I have 04h00 post on my vehicle. I wake back up after going to sleep after getting relieved at 05h00. Still windy, jamming to country music awaiting our next mounted patrol. 09h30, we start our movement north as soon as we set up another temporary operating post. I pass out. I forced myself to obtain the ability to fall asleep in any position with full armor on my body covered in sweat, salt, and dirt. I wake up again at noon sweating bullets and feeling like a piece of beef jerky that was hit by a train. I scrounged for water from the cooler. I pound two bottles down. Shortly after, LCPL Dunjay comes over to show me his pet lizard he caught and tied to a string. We provoked the lizard to bite our fingers and laughed and joked around just to keep ourselves entertained while we sat around some more. I then decided to listen to some more music on the iPod. I also just got done eating a halal lamb meal that one of the interpreters made and thought it wasn't half bad. At this point, anything was better than an MRE. 08h00, we closed off the coil, moving the vehicles together a little bit to keep a tighter camping circle. The sun is almost completely down. I break out pictures of loved ones and then lay my head to rest until it is time to be woken up for fire watch.

June 29 – Roving fire watch. All was quiet. I could hear nothing, for as we were in the vast desert so far away from anything. 07h00 wake up with intense pain starting from shoulders going down my entire back. For the past couple of days, I have been waking up incredibly angry with everyone and everything. I am slowly starting to go insane from this desert. Everything big or small pisses me off. We drive around in a terrible sandstorm for a few hours; visibility is incredibly low. We spotted, snapped, and apprehended two members of the Taliban transporting 15 bags each of urea/ammonium, which is an illegal, highly explosive substance. These members were driving a small tractor pulling a wagon filled with pillowcase-like bags. Each bag weighed about 100lbs. One of the men also had heroin in his man-dress. Man-dress is what we would call their over-shirts or what they would call *Perahan*. We often referred to it as a "man dress" because it was a long overgarment that resembled a dress. I conducted a physical examination of both men in the sandstorm. It's not that hot out today because the sand in the sky was blocking the sun. The body odor of the individuals we would catch was unlike any wretched odor I've ever smelt. The best description of this place is the hell scene in the 2005 movie Constantine. We waited for Crazy Train to come out to our position, but air was red, so we were under the impression that they might end up not coming until the following day. We were running scarce on water already. From 15h00 that afternoon, we stayed in a coil all night trying to shelter from the sandstorm while keeping a watch on the new detainees.

June 30 – 07h00 Woke up feeling good due to finally getting a full night's rest, which was extremely rare. We await resupply now that the sandstorm has stopped. We waited around doing nothing for three hours. 10h00 Crazy Train finally arrived to resupply us on water and take away the drugs and two detainees. The Marines decided they didn't want the tractor and wagon, so after they left, Gunny decided to start it and slam it in gear to ghost ride out into the desert. About 1 hour later, the tractor revealed itself from behind the sand and started doing circles around our coil. It was hilarious. Someone in the group spotted it thinking another person was off in the distance in another vehicle. 15h00 we move to a new place and set up an operating post. The breeze is flowing. I set up some shade and relaxed as I listened to 1st Sgt tell us stories of when he was a young gun. 19h00 still sitting in the coil talking with Loui O waiting for bedtime. When you are on deployment like this, you tend to talk about everything and anything. You get to know every detail about a person you operate with. You learn about their likes, their dislikes, any and all details about their past experiences with no filter on the story. Public non-judicial punishment (NJP) goes down for LCPL Wagoner falling asleep several times on watch. NJP was where a unit would make an example of an individual who needed discipline. This normally would be reading out of the offense and what the punishment would be for that offense, such as loss of rank or being assigned extra duty. It served as an incentive for others not to get in trouble. LCPL Powder Smash switches vehicles with LCPL Rod from Black-8 to Black-6 because apparently, MSGT picked him up and yanked him out of his driver's hole violently and injured him. I guess the desert frustration was getting to all of us. I fall asleep at 21h30. Bedtime was my favorite time of deployment. That and eating. The Marines would make fun of me for sleeping whenever I could. If it was one thing I learned from Iraq, it was that time goes by faster when you sleep and guess what, it sure as fuck did. I slept as much as I could whenever I gained the opportunity to. Part of the reason for that was because of how sleep deprived we always were from our operations out there. Another reason was because being out there doing what we were doing was extremely depressing and it makes you tired all of the time.

Chapter 8: Pressing On Through Hell

July 1 – 02h00-03h00 I stood roving security watch. I witnessed an explosion in the distance at 04h00. We then began movement to a new position. We drive around some more, searching for bad guys pretty much the rest of the day. We coiled up again for the night. I sat on the LAV, watching planes drop bombs in the distance and flares that would shoot off in the sky to my right. I'm not sure what they were engaged with, but I made a prayer and a cheer for whoever was fighting. I called Dad on the satellite phone, and he told me he had lost a lot of weight. It was good to hear his voice after what seemed like an eternity. We had to pack up quickly due to spotting a vehicle heading straight for our position. Black-6 and Black-4 rushed on to interdict and flank. Upon intercepting and investigating, there was nothing significant to find on the vehicle. 22h30; I bed down for the night. 23h00 woken up for watch for an hour, halfway through, the explosions stopped. I went back to sleep after trying to call a hometown friend of mine named Neha because I needed someone to talk to close to me back home, but no answer after 4 attempts. I go back to sleep only to be woken up again within the next hour to look at a BFT message sent from BAS about something I did not understand. This went on, getting woken up several times because the BFT was sent to every vehicle in my platoon, so at different times 5 different people woke me up to inform me of the same strange message.

July 2 – Wake up at 04h00 to go resupply from Crazy Train. SSGT Chantickle started the day by being a douchebag, thinking he knows everything and was saying some bullshit, simply annoying me as usual. Powdersmash ran into the resupply 7-ton and smacked his head. Not sure if he did it on purpose to get attention or to try to hurt himself and try to get sent home. No one could ever believe anything that came from that Marine. After resupply, we sat around in the coil sweaty, tired, and caked in dirt that covered our bodies from head to toe. 13h00, we linked up with the battalion CO and British forces to conduct mounted patrols. We drove around for a while, and dust and sand was piling on me again. Red-3 broke down, so we recovered them and coiled up and hung out with that platoon for a while. For the rest of the day, we just hung out with Red Platoon.

July 3 – Woke up feeling great due to a full night's rest. The company commander tasked our White Platoon to find 2 sheep for our 4th of July feast. We clean out the back of the pig as we jam out to music, sweating bullets. Noon rolls around; I eat lunch and do laundry in trash bags. It was 18h00. Berserker games begin for the 4th of July celebration, which consists of the kettle bell toss, LAV pull, and tug of war. It was a good day to boost morale and see the rest of the company. It was a break from hell within hell for a short while, if you understand my meaning.

July 4 – 03h00, we drive out 800 meters to collect air drop resupply, but the air drop never came. 05h00, we drive back to the company coil for rest. 08h00 I wake up to the hot sun beating on my face and the sound of Marines chasing lizards. Earlier this morning, I took doxycycline (doxy) without food, which is a huge mistake, and puked my brains out about 5 times. The battalion made taking doxy a requirement for malaria prevention. If it wasn't taken with food, it was known to cause an upset stomach. Those who couldn't take it had to take tetracycline every day as a substitute. I sat around playing cards. 15h00; part of HQ comes back to the coil from Camp Payne with steaks, a grill, melons, apples, Gatorade, soda, chicken, and hotdogs. We dug holes in the ground and lined them with trash bags. We then filled them with water and the drinks, started grilling, and had a feast. We hung out behind the pigs smoking cigars and talking about the good times. Our air strike lieutenant calls in 3 jets to fly real low overhead for us for our fireworks show. Because we were overseas, they could fly much lower than normal, and it felt like we could reach out and touch them as they whizzed past, fast beyond belief. We played Frisbee until 20h00 and started a night movement to a new position. It was a different Independence Day for us all, for sure.

July 5 – 04h30 We continue to move to a new position to interdict vehicles after 3 hours of sleep. We chase vehicles around the sand dunes for a while then resupply Red Platoon. Vehicles will spot us and run away, which means they either have illegal shit or not; that's pretty much the only action we have been seeing so far. We then drove to a different location, resupplying the rest of the platoons. The wind was violent today, which means yes, it's not hot as balls under shade but fuck there is sand in my eyes and in every single orifice of my body. Meanwhile, I have been treating one of our interpreters since the afternoon of the 4th for what I believe is appendicitis. The patient was stable, but I wanted to play it safe and get him to a higher echelon of care just to be sure. Fuck it; we sent him back to work until we go back to Camp Payne on the 11th. Guess if it actually was appendicitis, he would have been in excruciating pain and unable to operate. That obviously wasn't the case. I called Mom and Dad on the satellite phone when the platoon was passing it around. It was nice to hear their voices again. Not much I could tell them other than I was "Safe and doing fine." I couldn't have them worry sick about me while I was in hell.

July 6 – Slept in until 11h00, then drove around desert some more. White Platoon chased the truck that got away but left behind a green digital camo jacket, piece of flak, cell phone, pain pills, and a blanket. It wasn't strange that the Afghan people had these items, but we wanted to make sure they couldn't use them against us in any way. We inventoried it for evidence and then continued to drive around the desert. I can't wait to get out of here.

July 7 – 18h00 Literally nothing significant to talk about today. We drove around resupplying the company. I am sitting here wishing I were back in the states. All I could do was think about being somewhere else and look forward to getting back home. It was becoming sickening thinking that I was stuck here. What helped push me through was the fact that I had much pride in serving my country and doing something badass with my life. I enjoyed taking care of my Marines. It was the boredom and non-action days that really drove me insane.

July 8-9 – Drove to a new OP (operating post) way up on top of a big hill overlooking some towns. We got very little sleep that night due to an overnight movement.

July 10 – I worked out all morning. I had to send one of the Marines back to Payne with cellulitis on the left knee. Another Marine burnt his left hand on the vehicle muffler. Finally, things were happening. Unfortunately, things that happen and require me to work are normally not good. Today was an extremely hot day. I was sweating my balls off, even in the shade.

July 11 – 03h00 reveille. We packed up our sleeping gear and prepared for another movement. We conducted a 04h00 movement back to Payne for training. It was nice going back to base after a while being outside of the wire. Every time we went back, it seemed there was always something new about the base being built or improved. It was also nice to take showers and sleep on an actual mattress in air-conditioned tents. They had a morale welfare and recreation (MWR) tent there. A place where we were able to use the computers and talk on the phone. It was like the DMV there. You take a number and wait to get called for your turn to have 30 minutes of time to do whatever you need to do. 05h00; a 7-ton rear ends Black-7 on the vehicle ramp where we staged the vehicles for maintenance. Red takes on whiplash. The back hatches of Black-7 are destroyed as well as the front end of the refueler 7-ton. Marines rig it to another 7-ton for a tow. I can already tell it's going to be a long-ass, dreadful day. Nothing else significant happened after that incident for the rest of the day.

July 12-21 – Rest and refit, good chow, air-conditioned tents, mattress to sleep on, and MWR. Our mission while at Payne was to do just that. The higher-ups would do the same as well as plan and prepare for the next missions. My job was to restock on medical supplies, conduct sick-call for the Marines, and debrief the Chief. This went on until it was time to move out and operate outside the wire again.

July 22 – Movement out towards our mission point. During our movement, the vehicle commander has the job of navigating and assisting the vehicle driver with maneuvering obstacles. The vehicle commander sits propped up at the highest point of the vehicle with the best point of view. While on our movement on this particular day, I heard 1st Sgt scream, "Hit the brakes!" Without hesitation, Loui O did exactly what he was trained to do and slammed on the brakes causing the vehicle to come to a violent halt. I couldn't brace myself quick enough and lunged forward toward the opening lip of the back hatch of the vehicle. I was standing facing forward. The violent motion of the vehicle lifted me with all of my gear off my feet and headfirst into the lip of the roof hatch of the vehicle. I was wearing my helmet, but the force of inertia was so powerful that it caused my helmet to lift off of my forehead, exposing it to impact the metal hatch lip. I slammed my forehead hard enough to rattle the teeth in my head. According to the vehicle crew, the vehicle commander suddenly halted the vehicle due to the spotting of what distinctively looked like an IED (improvised explosive device) that was planted on the road. After the stop, everyone tried checking on each other over the radio, and I was unresponsive. They looked in the back of the vehicle to find me lying there with blood all over my face and head. I was unresponsive until the 1st SGT climbed back to shake me awake. I had blacked out from the impact. It was the first time in my life that I had ever lost consciousness from a head injury. When I came to, I remember seeing 1st SGT in front of me all blurry. I was experiencing a pounding headache. They called over one of my Corpsmen from another platoon to do what was called a MACE (military acute concussion evaluation) exam on me. I kept saying that I was fine because I didn't want to get sent back to base to sit around and do nothing. Plus, the thought of something happening to any of my Marines on this mission while I wasn't there would have destroyed me. I would never be able to forgive myself. I sucked it up and took some meds for the pain after getting cleaned up. We pressed on. My head continued to pound throughout the night as I tried to get some rest. [Little did I know then that ten years later, I would be fighting the VA (Veterans Affairs) for compensation for constant headaches and migraines that would persist in the area where I smashed my head on this fateful day.]

July 23 – Wagoner finds a huge scorpion. He caught it and placed it in an empty Pringles can, and we all looked at it with amazement. It was the most interesting thing that has happened all day. I still have a throbbing headache from smacking my melon and a nice slice on my forehead. The lack of detail written about this day should be enough to tell you that we did nothing but sit around in a platoon coil all day awaiting orders from higher up the chain of command.

July 24 – 01h00 – 02h00 2 tractors roll by that were already snapped by another platoon. I sat around all day until 15h00. That was when we were given the order to roll out to stage outside the town we are rolling into tomorrow. 16h00; there was a roving post uproar with me. [I'm going to backtrack a little bit to explain this situation. The last time we were resting and refitting at Camp Payne, I was debriefing my chief in the BAS (Battalion Aid Station). After that, and while catching up with some of the other Corpsmen, the Chief then overheard me talking aloud about an event I was sharing while on roving security post outside the wire. My Chief approached me and had a very serious conversation with me. He confirmed that I mentioned that I was standing roving security watch outside the wire. He then gave me the direct order to never stand that type of post again and had included that it violated the Geneva's Convention Code of Law due to the fact that Corpsmen were considered non-combatants. I understood and complied. I did, however, mention to the Chief that my Marines were not going to like what I had to say when the time came for them to place me on watch again. The Chief explained to me that there would be consequences if I disobeyed his direct order. I told him, "Roger that, Chief!" and he then gave me a paper copy of the order to give to whoever challenged my denial of standing post.]

Well, it was on this evening where that time came where I had to inform the Marine in charge of creating the watch bill to leave me out of roving security watch. He then informed the Master SGT and the 1st SGT, who then flipped their lids over the ordeal. I later overheard that they were communicating to each other that they were preparing to issue me non-judicial punishment (NJP) for dereliction of duty or failure to obey an order. So, you can see the dilemma I was in. It turned out I won that battle because they were forced to fact-check it with the company commander and had to back off. Man, were they pissed. Their retaliation method was to make fun of me for several months thereafter and continuously refer to me as "Geneva's Doc." It didn't faze me one bit. The satisfaction of standing up for myself, knowing I had my Navy Chief that had my back was satisfaction, and the more they made fun of me, the more I'd feed into it, replying, "That's right!" just to go along; with the charade... Assholes.

I was placed on a heavy dose of radio watch in one of the communications vehicles for the remainder of the deployment, which was fine because it wasn't illegal according to the law. I could tell the rest of the Marines were slightly confused and maybe annoyed, but I didn't want to feed into the frustration because they knew it wasn't smart to mess with someone who is in charge of saving their life if it came down to that. [In an ironic twist, but relevant instance; after deployment, one of the Corpsmen from a later unit who was in fact forced to stand gun watch was killed during his turn for post. Needless to say, I'm sure some of the Marines in charge of that unit were relieved of their duty.] That night, we could hear explosions in the distance at 19h00. 20h30; air spotted 4 Taliban activating IEDs outside of our position, we requested to drop bombs, but the higher-ups denied the request.

July 25 – I stood radio post feeling proud that I hadn't backed down from much higher-ranking Marines, but strange because I knew they were pissed at me. Reveille at 08h00; we sit around all day. We were anxiously waiting to get the order to begin our assault in Durzay. Sitting on radio watch 15h20, we hear machine-gun shots in the distance and see a cloud of smoke from a helicopter that laid the smack down on that part of the town. We find out the invasion gets pushed back 24 hours. Bring on another day of agonizing boredom.

July 26 – Sit around all day awaiting resupply. The company preps for combat by cleaning weapons and checking all equipment is operational. Everyone is still anxious about tonight's assault. 18h00, we get resupplied and link up with special attachments and explosives ordinance disposal (EOD) teams. 22h00, we rolled out to our staging point and bedded down for a couple of hours while EOD went to work on those IEDs that the Taliban activated the night of the 24th.

July 27 – 02h30 I stand radio watch and watch EOD roll back in. 07h00, we get radio checks and gear checks. I am so nervous and excited and ready for anything. I wasn't even concerned about dying. At least it was something to do other than sitting around. We roll out and begin our invasion on the city. Highlander calls to us while en route at 07h15, saying hold our positions due to a hazy dust storm. We hold and wait for the wind to die down. The anticipation is making everyone agitated more and bloodthirsty. I feel nauseous because Loui O farted, and it smells like death from his bowels, and I'm trapped in this closed up sweaty hot vehicle, not able to see what is going on around me because I did not have a periscope in the back of the vehicle where I operated. I was hoping and praying that the last smell before I died wasn't a fart. That would have been a shitty way to die. 09h00 air goes red due to the dust storm. Loui O and I reminisce with 1st Sgt about life. 13h00 still chilling and listening to music. I think about the special times, people, places in my life, and to myself, I think about and wonder if I was going to die today. The feelings flowed through me as if it didn't really bother me but only in the way where I knew it would hurt my loved ones if I didn't return home. 14h00, with no sign of the sandstorm dying down, we moved to another position to get rest, and the waiting game begins.

July 28 – Wind dies down a little bit, but air is still red, so the waiting game continues. Did nothing all fucking day except stress about this big mission and tried to rest.

July 29 – Word gets passed that our assault is going down tomorrow. Sitting around all day again listening to angry music thinking about this nightmare I had last night about ole what's her face. 19h00, we move out to an OP position. We take route 'bloodstripe' 500 meters from where a vehicle from Charlie Company hit an IED not too long ago. 21h00, we coil up just outside the city and observe all activity, i.e., men activating explosives and people going about their business. Through the scope, we notice a man performing intercourse with a donkey. That's right! There was an old Afghan farmer fucking a donkey. It struck everyone by surprise when all we could hear over the radio was, "Hey everyone! Traverse your turrets to this coordinate!" It was the most complacent out of this world, and I couldn't believe it was really happening in front of our eyes moment we've all ever shared together. We all recorded video and took pictures through the periscope that was night-vision green. [Yes, I still have the video, and yes, you can see it. I ended up showing everyone (friends and family) after nonchalantly throwing in the mix of how deployment was; I casually mentioned I have footage of a man fucking a donkey, and suddenly that's the only thing they care about; seeing haha!]

July 30 – Our assault finally begins. We push out, moving into the city. On our way in, we spot more personnel (3 individuals), and we monitor them, ready to blast the shit out of them. Team Thor pushes out the MARK BOT to assess for IEDs. The MARK BOT was a robot used to secure hazardous areas with possible IEDs before sending out a human to investigate further. We move in, surrounding the city, following Thor. People of the town start acting surprised and react by moving around, driving on motorcycles, and in cars observing our movement. There are a lot of people here. "To my family. If I die, I love you all, and I will be watching over you, smiling down as you move on with life. I knew what I was getting into when I signed up. I choose this; to serve my country and to make something of myself. If shit gets crazy and I don't make it out of here, I will meet you all in heaven." [Unfortunately, the good die young, so my ass is going nowhere.] EOD continues sweeping for bombs as we observe movement. I lay low on the bottom stretcher in the vehicle and passed out from exhaustion. Guess I was tired from being overexcited for several days and all the emotions that came from anticipating finally getting into a kinetic combat zone. Suddenly, we get rocket attacked! While being attacked, it was like I was sleeping through it, dreaming about dodging rockets in a school gym with my ex. Strange. I awoke several hours later at 11h00. 1st Sgt told me I slept through the barrage, which made me laugh; he thinks I am crazy. We sat around in a coil for a while, waiting for something to happen while we waited to talk to the town elder. The nice cool breeze carries the smell of rotting cow corpse into the vehicle. Lovely. We pass the time by cracking on each other. Morale had suddenly been boosted because we were all doing something exciting for a change. We get the order to dismount and help resupply the other platoons in the city with chow and water. We end up having to cross a water creek they use as an irrigation system for crops. The water was over our heads in some spots and filthy green. We get the thirty-plus boxes of chow and water across and have to walk another mile soaking wet following the path of red spray paint arrows that EOD swept for us. We make two trips and complete the task cussing our heads off along the way. Sun starts to set, and Muslim prayer is heard throughout the land. It was eerie to us visitors. The locals get on their hands and knees to pray. The people passing by our patrols look at us with disbelief. I can tell we are not welcome here. We cross the creek once more to mount back up on the vehicles, but on our way, the Marines behind me are yelling out to Rodriguez. I look behind me and see nothing but the top of his helmet out of the water. I quickly throw my gear and weapon to shore and dive in towards him and manage to pull his head above water and push him to shore while I stay completely submerged in the hole we had sunk into. He would have drowned if I didn't act quickly. As I stayed submerged under the nasty water, I used all of my strength to push the Marine with all of his gear as high as I could. I was holding my breath and walking through the mucky bottom carrying him toward the muddy bank. We eventually made it back to the vehicles soaking wet- covered in mud and nasty smelly water. If I die a horrible death from illness later in life, it will probably be due to being exposed to whatever the funk was within that water. We close all hatches for the night, and over the net, we hear radio battalion intercepting a cell phone conversation that is translated into "Those Americans have the ammo, but no balls to use it." Everyone suddenly grew more bloodthirsty. We bed down for the night, and I have this crazy dream I met Mom in Alaska to sell crack to Danny. Strange things are happening, and it is beginning to affect my subconscious mind while I slumber. But things are finally happening nonetheless.

July 31 – The night was silent. It seems like it was the calm before the storm. 04h30 reveille. We prep for combat and hear explosions go off to our right. Four explosions go off within the hour. We prep to go talk to the town elder, and air goes red. If a Marine goes down, it's up to me to keep him alive until we can casevac. MSGT asks Loui O and I to build a raft for the resupply across the canal. Sgt Weisser [the head mechanic] comes to our pig to kick it with me while we wait to receive more chow and water to resupply the platoons deep in the village. We receive the order to go patrol into the village. We cross the bridge and get the radio traffic of the bomb-sniffing dog Molly finding an IED charge. EOD begins their sweep. BOOM! Shrapnel is flying 100 feet in the air. The frightening feeling can only be described in so many words. Bone shaking. Instant panic. Everyone on patrol hits the deck. We wait for indirect fire and run to the scene as parts drop from the sky all around us. We do our best to maintain composure as we recover from the blast just behind us, vehicle parts falling from the sky and screaming from different directions all around us. Had there been an inclusion of indirect fire from the enemy upon us, we would have all been dead ducks in the water. My hands are shaking, and I am frightened. We arrive on scene but can't cross the creek due to another possible IED. I look over the sand berm next to us seeing a completely demolished EOD vehicle. Amazingly everyone escaped with minor injuries. We set security as the wounded get ground evacuated and hear several more explosions in the distance. We wait for several more hours until it is safe to move across the creek back to the vehicles. We link up with the rest of the platoon and head back to the coil. Everyone is still in shock about the events that took place on this day. We were still ready for whatever came our way. Exhausted and starving, I try to relax behind my LAV as the sun goes down. Thinking…we walked right on top of that IED today, God, I feel so lucky to be alive.

Chapter 9: Kinetic Encounters

August 1 – 04h00 reveille. Explosions continue throughout the morning. 11h00, we get resupplied on chow and water by Crazy Train. 12h30, another RPG attack occurs on patrol in the village along with an IED pipe bomb discovery in a haystack.

1600 multiple more IEDs are uncovered throughout the day, including a weapons cache filled with ammo and a 155mm shell.

August 2 – Another 04h00 reveille, explosions continue to go off. EOD continues to dispose of IEDs all morning. A patrol of Marines sets off to check out the weapons cache. It appears to be Russian. MSGT informs me that I need to do a special mission today in the village. The plan is after resupply to the platoons; my task is to cross the creek with special medical gear to assess and treat injuries of Marines that need medical attention. I choose my buddy LCPL Devin Brown to accompany me on my mission. 15h00 White Platoon reports contact of small arms fire from the north. We are on our way out to begin our mission. White takes cover and begins maneuvering to flank the enemy, but there is a lot of open ground. I am anxious to get out there off this vehicle and pound the ground and assist. Jets are overhead, flying low, searching for the enemy. I have Dev and Boslee by my side, ready to jump into action with me. 1st SGT yells for us to stay down. We can hear rounds going off over our heads and explosions all around us. We continue to stay low with nervousness and excitement. Composure and mission focus was key. I soon began to grow agitated that the enemy was shooting at us and had us pinned down. The gun fight continues for 30 minutes then dies down. The Marines are closing in on the enemy position. I continue to wait while listening in on the radio chatter. Black-5 gets position reports from each section and requests an air strike on the compound from which was delivering machine gun fire. We lay low, waiting for the platoons to reach a safe distance away from where air is going to drop bombs. While doing so, they try to avoid possible IED sights that air reconnaissance discovered last night. I finally dismount to cross the stream. I shimmy my medical gear across. The rope I was using as a guide snaps off the stick it was tied to, and I fall in. I surface and start struggling with all my combat gear onto the shore. I make it across and link up with the Marines, who exchange their stories of combat action with me while I assess their injuries. Every

Marine is covered from head to toe with flea bites. The sun starts to go down, and prayer goes off again. Still soaking wet and cold, I continue to treat the Marines. I notify MSGT that I am going to need a full resupply of medications, including insect repellant. I swim back over to load up on the pig, and we get resupplied. I stay soaking wet until I lay my head to rest.

August 3 – 07h00 Movement back to the village continues as I treat more Marines. Powder Smash continues to get his ass handed to him for screwing up. He came to me yesterday claiming he was going to commit suicide, so I dealt with it accordingly. He is now without his weapon and ammunition and is doing bitch work instead of joining us in the fight. What a time to have to deal with something like this! I was dealing with physical and mental medical treatment while dodging enemy fire and trying not to step on bombs. Everyone thinks he is pathetic and weak. I was convinced he was not cut out for this job at this time in his life. I did due diligence to be supportive in accordance with his needs. I still and always will care about him and his well-being as a person. We get the order to pack up and move back into the village. We get to the resupply point and cross the creek again. I shimmy across the rope to treat more Marines and to link up with my junior Corpsmen Doc Mantis and Doc Hermon. We discuss what medications and supplies from the rear they need, and I make a list. We say our goodbyes, and I head back across the creek on the rope again. I friction burned the hell out of my leg on it. We load up and head further down into the village and link up with the rest of the platoons. I get my medical gear out, and I do my rounds, checking on everyone in the area. The sand fleas were ruthless out here. The town elders came by to talk with the CO, and we waited for the next order. We moved back to the coil outside the village after mission completion and began weapons maintenance. 10h30; a man on a motorcycle drives super close to our position and circles us. We shoot pen flares at him to stop. One of our vehicles presses out to search him.

He poses no threat, but we tell him to stop being a fool and never do that again because he might be shot next time. I need a nap. I awake an hour later at 11h30 to Orta waking me up because the commanding general came to our coil for a visit. Jimmy Z comes to my pig for me to check out the rash on his back which was from sand fleas. Shortly after, Dunaj comes over for a venting session about the idiocy happening on his vehicle, Black-5, and to get a thorn lodged in his thumb removed. I removed it with a scalpel. At 12h30 we heard machine-gun fire in the village, everyone gets amped, and the radios went crazy. The firefight continues, and we call for fire from air after we get a grid of the enemy's location. Black-5 notifies everyone air will be on station in 15 minutes. The fighting continues. It feels like living the legend of the Marine who doesn't get to kill gets pissed and feels like a nobody when he watches all of his comrades go to battle in a blaze of glory, and he can't. But we are all a part of the fight regardless of whether we get shot at or get to kill someone. Being constantly around Marines has an effect on me as well. I, at some points, found myself growing bloodthirsty but soon snapped back to reality to remind myself that I am a lifesaver, not a life taker. I like it better that way. The fighting and gunshots continue for another hour before it dies down and then air arrives on station. The sound of jets and helicopters flying low and gunfire reminds me of every war movie I have ever seen. This was real life, but it felt as if I was in a movie. Shit was getting more real, faster, and louder. The enemy shifts its position, and the fighting picks up again. I am staying low, fully ready to jump into action as soon as a Marine needed that higher echelon of medical care. 25mm cannons go off from the LAVs and blast the shit out of the buildings the enemy is in. 14h00 Thor 3/1 EOD reports hitting an IED. No injuries, but they have sustained vehicle damage. Smoke billowing from the building that is on fire from our 25mm hit fills the air. Helicopters spot personnel fleeing from the burning structures. The enemy is on the run. My buddy Boslee, the platoon mechanic, comes to my pig and kicks it with me for a while. The fighting has died down again. I take another nap and wake up several hours later, sweating like a pig. We get the order to resupply Blue platoon. I make the call to take LCPL Webby out of the fight to routine cas-evac him due to an uncontrollable growth on the left side of his mandibular angle going up behind his ear. We watch 4 individuals through our thermals dig holes and place objects in the ground that night.

August 4 – We cas-evac Webby to the medical officer in the rear at 05h00. I receive antibiotics to administer if the patients' signs and symptoms increase. Also, get my ass handed to me for not enforcing hygiene to the Marines, which was laughable because it was near to impossible to stay clean in the situation we were in out there. We return to the forward coil and wait out for the next mission. I am sitting here chilling, listening to music, wishing I were surfing in California with my friends. We try to provoke the enemy out of hiding. Patrols continue throughout the day. 16h00, 1st SGT yells for me to load up. In a hurry, I get in the vehicle with no idea as to what is going on or what is about to happen. Black-6, Black-4, Black-8, and EOD reach the inner limits of the village where the other platoons are providing night patrols and overwatch. My heart is racing, and my knees are shaking because I still have no idea what we are doing. We park behind some mud huts on the riverbank and get out to observe the strange activities from personnel inside the village. The company CO gets on the net and debriefs everyone on today's progress and the upcoming events. There will be a jihad (holy war) tomorrow, and we are the main attraction in town. We prep for the night and prepare to move to an offensive position. 21h00, we mount up and start our movement to this position. Lead Vic Black-4 gets stuck in deep mud, and we attempt to tow them out, but the anti-IED "Up" armor kit was acting as a spade digging deeper as we pulled. Marines dismount to dig.

The CO decides for certain members, including my crew Black-8 to hold firm here for the night while the rest of our team presses forward on foot to take the offensive position that we sought out to achieve with the vehicles earlier. We set up a position for cover and hold firm. This is where we will be staying until sunrise. I can already feel the sand fleas under my flak gnawing away at my sweat and dirt-covered skin. We are in a very dangerous position tonight. I feel scared and excited to get some more action. Sand fleas bite the shit out of me throughout the night.

August 5 – 03h30 reveille. I hardly got any sleep. We tac up and start our dismounted patrol out to a large hilltop between all 3 villages. Derzay, Zar Banadar, and Safar Bazaar. We reach the top at 04h30. Everyone in town knows we are here. EOD uses the minesweepers and finds a 25lb IED on the south side of this hill. 08h30, as we continue to monitor the towns, a loud explosion goes off. Explosives were detonated in one of the towns real close.

We communicate with each other on the radios and continue to lay low and observe. 11h30 I am sitting in a hole in the ground next to Wagoner, the radio operator who took the following picture of me.

The sun is beating down on us. Radio Battalion rogers up to us that they intercepted a cell phone conversation that they are planning to attack soon. We continue to observe the villages while praying to God for some clouds to block the sun. 12h45 Doc Rayman radios up to me about one of his Marines in White Platoon, Corporal Walshart, getting caught 5 meters from an IED blast. He tells me he scored stable on the MACE exam but was in pretty rough shape, and he wanted to know if there was anything else he needed to do. I told him to send the report via BFT to the medical officer. He said thanks, and I replied, "Roger, Thundercat, Care Bear out." At that moment, I knew he was on the other end laughing his ass off.

The sun is still beating down on us, and nothing is going on in the villages below. I ask Wagoner how he is holding up, and he states that "he is losing his mind" yep, that sounds about right. 13h00 explosions and gunshots right below us in the village to the north. These sounds and the reality of actually being here make my bones shake. The explosions and fighting continued for an hour until it ended with an enormous boom from HIMARS, which were rockets from our base about 100 km away. The sounds are silenced. A giant mushroom cloud of dust and smoke rises to the atmosphere. Shutting the enemy down with massive firepower never felt so good. Prayer is played on the town's loudspeakers. 15 minutes of silence pass, and more gunshots are heard from the east. It seems we are surrounded by horrifying sounds of war. I am still sitting in this hole. I grab ahold of my necklace pendant that the Bierstos gave me and begin to pray. I have never been more frightened in my entire life. Silence goes about the land. Wagoner notices me shaking in the hands and gives me a cigarette. Without hesitation, I light it up and try to stay relaxed.

Songs of Muslim prayer are non-stop today. 15h20, another explosion erupts, and another firefight breaks out in the northside. I could see Black-7 rushing to a unit's position from the hilltop, which means we took another casualty because Black-8, my vehicle, is the primary cas-evac vehicle and Black-7 is the secondary. Jets fly by low overhead and shoot flares at the combat position. Jimmy Z comes to my side of the hill, and we talk about going home. 17h00, the CO makes the decision to spend the rest of our operation days out on this hilltop. We prep for the night and do a patrol to the pigs 600 meters east. We get to the pigs and load up on chow, ammo, and water. Night falls, and we head back to the hilltop. The patrol is very painful due to the heavy packs on our backs. 20h30, we reach the top and set up a watch roster. I stand the first watch on the south side. Sand fleas begin gnawing at my flesh. I am relieved by Orta an hour and a half later. I remove my flak and cover my exposed skin with gold bond foot powder to keep the fleas off me. I lay on my gear soaked with sweat and tuck my head into my shirt and don my gloves to keep the fleas off of me.

August 6 – 04h30 reveille. I awoke feeling like I got hit by a train with irritating red marks all over my skin due to sleeping through the fleas feasting on me through the night. I don my PPE and stand watch. An hour and a half later, I dig an area for the Marines to go to the bathroom. Then I dig a trash pit. My back is killing me. 08h00, the village people are going about their business on their crops and cattle. The town elders come to us to complain that we are trespassing on sacred grounds. Apparently, this hill we are on was an ancient castle that holy people lived in. There are ruins and remnants of room structure here. But it's the most advantageous position to be entrenched in the area, so we kindly tell them that they can have it back when we are done. 16h00, we spot 3 men and a donkey with jugs. The men began to dig in the ground on the route we traveled in on and started placing the jugs in the holes. The CO gave us the order to 'light em' up.' Corporal King and I jumped on the M240 machine gun and fired away. As we threw rounds down at the enemy, the dust began to rise in the air at the enemy's position. Loui O ran back and forth, supplying us with more ammo. He then yelled out, "They're running! Shift to the trees!" We anticipated their movement and fired some more. The spent brass flew from the ejection port of the heavy machine gun and pooled around us in our pit. Together we screamed out, "Get some!" and "Die, motherfucker, die!"

I killed 3 men and a donkey today.

The event was epic, and I captured it all on video. My heart was pounding, and my entire body was shaking. The thought of actually ending the lives of several human beings was giving me mixed emotions of excitement, pride, and regret. [Later in life, this would impact me in ways that require professional help to manage.] We collected the brass and received the order to pack up and move back to the rest of the platoons. We stayed the night out here awaiting an attack back. The patrol back was silent and uneventful. Our wits were high, and nerves were shot. It was a day that I will carry on my conscience and in my memory forever.

[Why did this happen? This is me paraphrasing paragraphs from cognitive therapy sessions with the VA ten years later are as follows.] {"Because there are evil forces in this world, and they must be stopped. Because those evil forces were out to harm those who had a different perspective on life. Their intentions were to kill us, so we had to kill them first. What I think about that day makes me wonder; what if they were just simple farmers. What if they were loving parents or guardians of children trying to feed their loved ones. The geographical distance was too far to know for sure what exactly they were up to. Suspicious activity in direct line of our operations was a threat to us. Suspicious activity qualifies as a hostile act or a hostile intent. Our mission was to eliminate all hostile threats. Even if this meant potentially threatening. This was a message to all other threats or even those thinking about fighting us. We weren't messing around. Life or death was a serious matter. We had already lost so many in this war. Tolerance wasn't an option.

I never once thought that I would take part in actually murdering another human, two for that matter. It was exhilarating. The kind of feeling you get when you catch the biggest fish of your life or harvest the largest deer you have ever seen. The kind of shaky with adrenaline that would disable anyone from being able to even hold a bottle of water or even a simple conversation.

It also made me have second thoughts about myself right there in that moment. I had just been directly responsible for ending the life of another. What have I done? What if they were innocent? Does God approve of this? Yes, the unit I was operating with did, and yes, America may, but what if the case wasn't that they were actually an evil force? Doubtful emotions began to flood my spirit. Had I now become an evil being in this world? Feelings of guilt rushed through me. I felt as if I didn't deserve to live anymore. Why did this happen? Why was I so eager and excited to do such a thing? Have I gone mad? I was disgusted with myself. I was also proud that I had a nation backing my action. I felt safe knowing that a higher power that gave me the order but unsafe knowing that an even higher power in the heavens was frowning on it all. I began to pray. Praying wasn't making me feel any better at that moment. Forgive me for I have sinned. I lost trust in myself. My esteem was haywire. I was proud to say yeah, I killed people. Cool right? What does that actually do to a man? It diminishes his intimacy in ways that are difficult to explain. Following years of nightmares, disbelief, guilt, sorrow, pride, and all the stages of denial, anger, bargaining, depression, acceptance (DABDA) have been my repayment."}

August 7 – I wake up several times itching from the sand, fleas, and nightmares. We clean out the pig and rest up while EOD preps to demolish every bridge in the town to limit the movement of the Taliban so they can't escape town. I head over to kick it with my good buddy Tien in his turret, and we share stories of combat together. I get the idea to go show 1st SGT Tien's amazing ability to glow stick. I call home and try to talk to family as much as possible. After conversing with home, I felt so happy. I couldn't wait to see them again. I go hang out with Hernandez, the bomb dog handler and play with Molly and go for a swim with her in the creek after Tien gives me a "Super moto" haircut which is comically short and perfectly within regulations. I even went fishing today. There are freshwater crabs, frogs, and tons of fish in this creek. 16h30, we move out of the village to link up with Jump Force and the British Special Forces. We coil up, and the force leaders get together and plan a special mission. MSGT calls me over to look at LCPL Terrely's left ankle. He was stepping off of the turret on top of the vehicle and landed on it the wrong way. His ankle was covered with edema and erythema, and dark bruising. We took him in our pig so I could continue to monitor and treat him. We standby and anxiously await the order to push out and begin our new mission. I go on watch at 23h00 and stand an hour then go to bed. I was really happy to see brother Dev again. We chatted about what we had seen today and about going home and surfing. I then slipped off to bed.

August 8 – 02h30 reveille. Not feeling good at all. I felt like I got hit by another train. I was so tired. We move into Bravo 2-1 position back behind the mud huts with the other platoons after we drop off Terelly to Black-7. We were out here now with the British Royal Marine Commandos, and we are doing patrols with them so they can occupy this area and so we can move out back to Camp Payne. The Royal Marines have different rules of engagement than us out here. They shot at the villagers just for gathering for a meeting today. I am sitting here, ready to jump into action if we take any casualties. 17h00, we displace to the H+S company coil into the rear in the desert to meet up with the Battalion Sgt Major and talk about Powder Smashs' malingering attempts. He compliments my high-standard moto haircut and tells me it's sexy. To be complimented by someone that high up is a rare honor. We coil up with the rest of headquarters and set up a watch list, and go to sleep.

August 9 – Wake up at 05h00 to the sounds of mortars exploding. Everyone is getting haircuts and making sure our uniforms are squared away and clean shaves. The Commanding General is coming for a visit. I wrap up LCPL Terrelly's ankle and administer medications. Jimmy Z comes to kick it with me in my pig while we wait for Major General Mills to show up. He never shows. We head to Camp Payne for rest and refit. I have a feeling I am going to be very busy seeing a lot of patients from this operation. Constant patrolling is wearing the Marines down; the men act tough with shrugs and a "Good to go Doc!" but I could see the weariness and the toll it was putting on them.

August 13 – The past 4 days, I have been way too busy to sit down and write in my journal. It's been as I have expected, with almost every Marine in the company coming into sick call for minor injuries they have sustained while out in the field, along with every Marine getting their sand flea bites documented in their medical records. Almost half of the company, including myself, also came to the BAS for a MACE exam due to being at least 50 meters from one or more IED blasts. LCPL Davis has been coming to the clinic with the exact symptoms of malaria, so I had asked him if he was taking his Doxycycline. He stated he was, so we had to diagnose him with a viral URI and prescribed him medications and 24 hours of bed rest. He returns for a follow-up and is running a higher fever of 103.9. His symptoms were increasing. I explained to him that he needed to inform us if he actually had been taking it. We sent him to shock trauma platoon (STP) to test, and he came up negative, but that general test had been sitting in the sun for God knows how long, so they tested him for each specific type of malaria, and sure enough, he came up positive for the worst kind. We sent him to another base several miles away for further testing. Meanwhile, the Bravo Company commander requests my presence in the company office, where he complained about how I was going to get someone into trouble for doing what I did to properly diagnose. I told him that I was just doing my job in order to save a Marines' life. I pretty much told him off in front of the rest of the higher-ups in a pissed-off manner. It seemed to me that he was worried about getting in trouble from his higher-ups because his Marines were not taking their meds which means that he wasn't doing his job to enforce that order. I had lost all respect for that

Marine that day. I then went to the battalion chain of command (COC) to speak to the judge and write a statement about this whole incident. I am currently awaiting the news on the results of LCPL Davis.

August 14 – 11h00 1st SGT called me into the company office to write statements on how LCPL Powder Smash was malingering since he came to our unit. Later in the evening, I went to the BAS, and HM2 Delacruz hands me papers to get started on Combat Meritorious Promotions for all of my corpsmen, including myself. 2 days went by, and there was nothing significant to write about.

August 17 – Last night, we got the order that we need to prep for another operation. 05h30 reveille. We pack up the pigs and head out with Blue platoon towards our mission objective. 3 hours into our movement, Gunny Navings' pig breaks down, and he decides he wants to unscrew the radiator cap… Without any gloves… I get the call over the radio that Gunny burnt the flesh of his hand, and I needed to go care for him. I packed some gauze and burn gel and rushed over to him. Doc Mantis was first on scene, and we both looked at his right hand. He burnt all 4 of his fingers that were already forming huge blisters all over. I make the call to cas-evac him via LAV back to Payne. Noon, we arrive at Payne, and I walk Gunny to the BAS, where the medical officer (MO) and I rewrap his hand with burn cream and dry gauze and give him light limited duty (LLD) for 3 days. 14h00, we push back out to link up with the rest of the Marines outside of our area of operation (AO). On the way, we see beautiful huge mountains and landscapes. I felt as if we were on a different planet. Mars would be a good description, I guess. After about 8 hours of driving, we finally meet up with the rest of HQ and bed down for the night outside of an uncharted village.

August 18 – 05h00 reveille. The sun comes up and reveals the towns we are about to patrol. In the distance, I can see what appears to be a huge castle structure that must be thousands of years old—surrounding it at a distance where the small square living structures appeared to have been constructed from mud and sticks. I have never seen anything like it before in my life. The Marines prepare for patrol in the towns surrounding this huge castle, and I hang out on Black-8 in case we have to casevac anyone. We sit around in the coil in the hot sun for several hours doing nothing until the company commander decides it is time to push back to Camp Payne. 15h00, we begin our 6-hour long movement. At 20h30, we reach the outskirts of COP Celtic/South Station, the forward patrol base on the other side of the river from Camp Payne. Black-4 has been running on a flat tire for a few miles, and it starts to smell like burnt rubber through the convoy. We pull into South Station to refuel and change out the burning flat tire before we push any further. As we pull in, we hear a large explosion about 100 meters in front of us. We looked around with confusion, and at first, we thought we were being attacked with mortar fire. We continue to move to the refuel point, and as I ground guide, an Afghan dog gallops over to me, and I put my hands out, thinking I hope he is friendly or else I am going to have to punch him in the face. He licks my hand and follows me with his nose up my ass. It is our turn to refuel. Soon after, we stage the pig behind Black-6, and we heard a 2nd large explosion just on the other side of the berm. Radio traffic begins, and we find out the explosion we heard before was an IED and that there are IEDs planted everywhere around us. Charlie Company Red Platoon got hit, 1 killed in action (KIA), and 10 routine casualties, one with severe damage to his eyes. If it hadn't been for us having to refuel and change a flat tire, it would have been my platoon that got hit. Orta finds out that this Marine that just died was one of his close friends. It was very hard for him to deal with the pain for several days, and you could tell he was upset. We sent a vehicle out to assist and recover the vehicle that got blasted. Since the route was crowded with EOD searching the area and Charlie Company doing recovery operations, we decided to hold it in place at South Station. Dev comes over to kick it with me, and the rest of the Marines bed down for the night.

August 19 – 04h30 reveille. We push out to do route clearance for Charlie, then head back towards Payne. On the road leading to the crossing point at the river, there were 2 giant holes where the explosion went off. Each was about 5 feet deep, with scattered shrapnel and debris everywhere. We safely make it across the muddy river and back to Payne. As we unload and clean the pigs out, Doc Trujillo walks up to me, and I ask him if he is okay. He informed me with every detail about what happened last night because he was the Corpsman on scene. "We took 10 routine casualties and one KIA," said Trujillo. I could tell he was shaken from the experience. Shortly after, Sgt Bananablak, my old vehicle commander when I was in Iraq, came up to me and explained everything that had happened through his eyes. He was also on one of the vehicles that got hit last night. He said the blast was so intense that it ripped all of his PPE completely off. I really liked Bananablak. He was a pleasant, kind, and soft-spoken fellow also from New York. He was mentally torn about what had happened. I did everything I could to be a good listener and provide my sympathy and care for his experience. The entire unit was all mentally torn about this tragic event. It is a pain shared by all. 11h30, we get done with cleaning the pigs, and I head to chow, where I cross paths with Navy Religious Program Specialist (RP2); he told me the KIA Marine was missing the lower left side of his body when they carried the corpse to STP. The rest of the Marines are being evaluated with MACE reports at the BAS. I finished chow and showered off, and tried to relax on my rack, still thinking about how lucky myself and my Marines in my platoon were. Tien comes into my tent to vent about his ridiculous vehicle commander (VC) SGT Lostass and all of the bullshit the rest of the crew on his vehicle had to go through because of the VC. We were all just doing our best to keep it together. Shortly thereafter, I tried to catch some sleep. It was becoming harder to have normal dreams considering what we were all experiencing. I was beginning to experience survivor's guilt. [Little did I know then that it would even affect me to this day, ten years later while writing this book. Tears still fall from my eyes for the loved ones of the Marine lost on this day.]

August 20 – I skate around the base until noon chow, then I receive three letters from different friends. Skate was a phrase that we would use in the military depicting the meaning of doing things other than labor or unenjoyable work. I received letters from my family. I wrote back to them later on after dinner. I had an awesome conversation with an interesting special someone back on the home front over the satellite phone. I went to the gym afterword's and then Tien did an awesome light show for some of us by swinging and twirling chemlights (glowsticks) from 550 cord. Another day has gone by, and I still can't wait to get back home.

August 21 – The day started off at 08h00 with taking LCPL Garl, the gunner from Blue platoon, up to the BAS for his follow-up appointment for the staph infection on his face. They were busy all night at the BAS doing exams on detainees and surgery on Doc Winter's ass. I agreed to help out with seeing patients so they could catch up on some sleep. Doc Hermon and Doc Rayman show up, and I teach them FMF knowledge from the book. I then took a nap out of boredom. Then I went to evening chow. I decide to order eye glasses for everyone in the company who needs them Guess I will do that tomorrow. I went down to the M Dubs (MWR) with Dev for a little while, then talked to Rayman about the glasses thing and went to bed.

August 22 – I woke up at 06h00, went to the BAS, fell asleep in the corner on the floor, and no one could find me until 10h00. Haha. I ordered everyone's glasses and then went to lunch. I finally got a chance to write to a girl I was talking to in a letter today. The day went by slowly, and we got word about doing an operation tomorrow that's going to last 6 days. I talked to LCPL Powder Smash on Facebook today. He said the psychiatrist has him on anti-depressant pills, and he is still in Camp Leatherneck. 19h45 I took a shower and packed my things for the field.

August 23 – We bring our gear to the pigs and get radio checks at 06h00. I am not feeling good at all. I have a sore throat, and my body aches. There's a dark sky-colored tan today. Air status is red due to the massive sandstorm. All operations are canceled until the storm passes. I go back to the hooch (word for tent, sleeping quarters) and sleep all day. I woke up for evening chow and then checked to see if the phone center was open. It's not! I guess we are still in River City status. River City status was when all communications to the outside world were shut down. The reason for this was due to the death of one of our Marines recently, and the battalion wanted to ensure that the family was informed properly. Wagoner informs me we were actually in River City status because a Marine from Echo Company got shot in the chest yesterday. I medicate myself and go back to sleep all night.

August 24 – Wake up feeling a little better but still shitty. Sore throat still and slight body aches. We go to the pigs and do crypto change over, which was resetting the radios, always a huge clusterfuckofagoddamnshitshow. Air was still red. I got back to the hooch and passed out. I wake up at noon, go to chow, still feeling like shit, self-medicate, and then pass back out.

August 25 – I wake up at 10h00 feeling well-rested and much better. It seemed the medications were finally working. The sandstorm was also clearing up. I hung out at the BAS all day relaxing and updating HM2 Delacruz and Chief on what was going on in the company. I watched the movie "Kick Ass" then went to chow. Worked out, then showered, which was always a moment of absolute bliss after weeks of not showering, then off to bed. The word was that we were most likely going out tomorrow for that mission we had been waiting on.

August 26 – Reveille at 05h00. I head to the ramp after hygiene, only to sit there on my vehicle for an hour before anybody shows up. I guess the word had changed, and I was not around last night to receive it. This morning was very annoying. Air was green. So, we were okay to start our mission, but since there was a sandstorm that lasted a few days, air was backed up on transporting overdue supplies. Everyone goes back to the tents to await if we can really start the mission. About an hour later, we get word to head back down to the pigs, only to be sent away again due to the confusion the higher-ups were having trying to determine whether or not we could start the mission. Meanwhile, I get to the computer/phone center where Sam, Powder Smashs' ex-wife, messages me saying that he told her he was in the wounded warrior's program for receiving shrapnel wounds to the back, which was not true at all. He had to leave us due to being mentally unfit for the job, pretty much-avoiding anything that was even remotely dangerous or difficult the entire time before being sent back to the states as a burden. I lost my mind again in frustration and told her the truth and then told every Marine in our company the blasphemy that Powder Smash was saying. Everyone now wants to beat the living shit out of him. Our hearts were still heavy with the losses and wounds our fellow warriors just endured and to have him lie in such a way is unforgivable. 13h00 rolls around, and we get called back to the ramp again. I am starting to lose my patience with whether we are going to start this mission or not. We get to the pigs and get ready to load up. We don't go yet. It turns out Wagoner lost a radio, and for the next 2 hours, the entire company was searching every square inch of the base, looking for it. 15h30 we roll out but can't go too far from base. It turns out all of Afghanistan has run out of hospital beds due to all of the casualties in the past few months. To minimize any more casualties, we can only go just outside the base until more hospital beds open up. We set up a coil and stop tractors on the road trying to pass by. The good thing about this operation is I have Molly the bomb dog and her handler Mike Hernandez with me, keeping me company. The sun sets, and we sit around waiting for more vehicles to interdict.

August 27 – 02h30, the XO wakes me up for radio watch. I am sitting there and have a runny nose and keep sneezing due to congestion getting worse from breathing in all the dust on our movement out here. The dust in this place was terrible, and there was no way to avoid it. The backdraft from the LAVs as they rolled on would send the dust pouring into the back scout hatches like a waterfall. An hour goes by, and Molly jumps in to keep me company and she keeps sniffing around in the dark and pulls out a thing of tissues and drops them on my lap. I felt so loved and impressed that she knew exactly what I needed. Good dog, Molly. Molly jumps out and snoops around the coil. 04h00 I get relieved by 1st Sgt, then crawl back into bed. Molly jumps on me to cuddle and keep me warm. I've made my decision about which kind of dog I was going to get when I got back. 10h00, I awake to the sound of Thor 3/1 linking up with us at our coil. Shortly after White Platoon links up with us as well. We push out to patrol and clear routes. We drive for about 4 hours, getting glazed with dust and sweat that turns into mud, covering me completely until I start to recognize my surroundings and realize we are going back to the area with the huge castle we were at on our last mission. We reach our destination at 16h30 after stopping and searching several vehicles along the way. We are directly behind a small village near the castle. Also, we were right next to a grave site. The village people can see our dust clouds behind the hills that we parked on, and they curiously observed as if they saw spaceships. Exactly like Iraqis, the Afghan people lay their dead on the surface of the ground and covered the corpses with dirt and rocks, then put sticks in the ground with rags tied on them. The scouts on Black-6 and Black-4 dismount to foot patrol into the village and interact with the people. We finished up and headed back to the HQ coil. We stayed there for the rest of the night.

August 28 – 03h45 to 06h00 I stand watch. I sound reveille for my vehicle. We sit around until 09h00. It is already getting hot. We drive around clearing routes alongside the river. 10h45 we go down a very steep slope to check out this small, abandoned castle made of mud on the riverbank. I struggle to find shade. It is uncomfortably hot, and I am starting to get agitated. We move further south along the riverside and find another structure made of mud bricks. Scouts dismount to explore the site, and it is beautiful. There are green plants, and we are right on the edge of the riverbank, watching fish jump as the water flows. It reminds me of good times back home. Unfortunately, the sun is directly above my head, beating down on me, and I am sweating like a hog with full armor on. We continue stopping at several small villages. The last village we stop at is at 16h00, and while hanging out in the vehicle talking with Hernandez, a young boy on a motorcycle rides up behind us. Orta, Hernandez, Molly, and I punch out to search him. He poses no threat and is cooperative. I searched the small bag he was carrying, and he had brand new shoes still in the plastic wrapping. Molly sniffs for any explosives on the motorcycle and finds none. I smile at the boy and direct him to drive around us. We continue to chill on the pig, sweating our balls off for another hour, waiting for the scouts to get done with their patrol. We drive off and coil up for the night.

August 29 – 05h00 reveille. The movement back to Camp Payne. We arrive back just in time for lunch. Work on vehicles blowing the dust off, washing at wash rack (muddy fun), and maintenance. We get done in time for evening chow. I shower and then debrief HM2 Delacruz on our mission and receive any word I need to pass to my men. Powder Smash was now transferred to Germany on his way home and possibly on his way to separation from the Corps. I focused on getting word passed to my junior Corpsmen and following up on their progress with getting FMF qualified and their letters in for the Combat Military Appreciation Process awards. I go to the computer center, check some emails, and read the news on current world events. Apparently, my unit company (Bravo Company) made the news a bunch of times for our work out here. Nobody decided to tell us about it though. Time for bed.

August 30 – I go down to the ramp after chow and re-label trauma medical bags on the pigs and help out with whatever I can on maintenance, and then head to the BAS to get resupplied on medications and ace wraps for my Marines. Pretty much the rest of my day has been spent running around medicating and treating my Marines. Later, I stopped by the junior HQ Marines' tent to find LCPL Hope getting tied up with duct tape. Apparently, he paid 2 Marines to hit each other in the balls, but to Hopes' surprise, they figured out the scheme and turned on him, encouraging the rest of the Marines to kidnap him and tape his hands and feet together like a calf at a rodeo. The Marines would roughhouse like children often, which was always fun until someone got hurt, which coincidentally was often. It is time to catch some Zs.

August 31 – I wake feeling tired as hell. Last night, several hours after I fell asleep, Devin came in and woke me up to patch his left ring finger. I cannot remember what he said how he did it, but the tip of his finger was chopped off. Not as bad as it sounds, but I patched him off then slipped back to bed; afterword's I went to the BAS to see what was going on in the Navy world and was dragged up to a MACE class at the STP. The class lasted until lunch. At the BAS this morning, I discovered the return of LCPL Davis. He had been receiving treatment for his malaria-like symptoms for several weeks. It turns out he didn't have malaria and has been deemed fit for full duty by medical. I go to chow with Wagoner, and he tells me what the Marines were up to today- 25mm firing range for them all day. Also, that there was a possibility that we might go on another month-long operation. I feel mutual about the situation. The boots decide to tie up the new guy tonight and hang him on the wall crucifixion style. I got bored and decided to hit the gym. Shortly afterward, I head back to the tent and meet King, and we decide to hunt for camel spiders outside. No luck. King then pulls out some mouse traps, and we bait them with combos, crackers, and peanut butter set them next to holes in the ground under our tent. I laid in my rack and read 2 more chapters of Tom Clancy's Splinter Cell and got some shut-eye.

Chapter 10: Wake Me Up When This Nightmare Ends

September 1 – Glad it's a new month. King and I discover catching a mouse in our trap. I head to the vehicles to see if there is any work I can help with. Later on, in the day, I discovered Doc Hermon and Doc Rayman took the FMF board, passed, and got their pins. Now it is time to motivate Doc Mantis to do the same. I stayed awake all night until at least 03h00. I couldn't sleep for some reason; I couldn't stop thinking about random things that kept my blood pressure on a rise as if I were constantly on the verge of action.

September 2 – Word/rumor gets passed amongst the Marines that we are going on an operation tomorrow for over a week. Orta gets meritoriously promoted to corporal today. I relax and wait for word to be passed about our operation. Word gets passed, and I hit the rack.

September 3 – 06h00 We start our 11-day operation. We drive out into the desert and conduct interdiction all day. Molly and Hernandez are with me on the pig for this operation, and so far, we have found nothing significant as the sun sets and we coil up for the night. I stand 1st watch, and then slip off into bed.

September 4 – 06h00 reveille. We plan to stay in this spot for the entire day. We must have parked on top of a nesting area because there are hundreds of flies everywhere. It is so annoying. 10h00, 2 trucks drive near our position. Black-6 goes to search for the one with sheep on it. The 2nd one has 3 men, 1 woman, and 3 children and was packed with jugs on the back. Hernandez, Molly, and I dismount to go help with the 2nd truck. Molly gets no hits for smelling any explosives, but when Sgt Abelar and I are searching, we are convinced we smell marijuana and make the driver unload all the jugs, and we search more, finding a 2nd hidden compartment and still find nothing. We eventually figured out that they had pot previously, but it's just the smell that remained. We let them go and waited for more vehicles to pass through. Sitting around until 1700 Black-6 radio comm goes down, so we load up and travel 3 clicks east to the Black-5 coil. I was happy to see Dev, and we talked about going to Florida. We sit around for about 2 hours while Black-6 gets their comm fixed. I decided to chow down on an uncooked can of green beans that mom and dad had sent in a care package. I have a feeling if the rest of this operation keeps going the way it has been, I am going to be bored as soon as I finish my Tom Clancy book.

September 5 – 06h00 reveille. The morning was pretty boring, and there were no vehicles to interdict. We packed up at 14h00 and resupplied on water and fuel. We drive for hours, getting covered in dust. 19h00, we stopped at the side of a mountain and coiled up with EOD and the rest of the HQs platoon, set up an observation post (OP) on the top with scouts and set up camp for the night.

September 6 – 06h00 reveille. It is cold this morning due to the wind chill coming from the hilltops. Our observation post scouts report a white truck heading in our direction. Everyone scrambles to get out of their sleeping bags, but as soon as the truck sees our pin flares, it darts off away from us and gets away. Another SUV darts by shortly after, but we were ready for him. Black-4 is already set up to intercept, and Black-6 darts off to flank. The rest of us are still on the hilltop watching them interdict, and I am taking care of Molly while Mike is doing OP watch on the mountain top. We decided to reposition every vehicle spread out along our interdiction route hidden behind hilltops separate from each other. From here, we have a nice overlooking view of the river valley and a better position to catch vehicles that try to pass through. It was 08h00, and I decided to cuddle up with Molly and take a nap. I awake at 11h00 pondering about the crazy dream I just had about being out here and fighting the Taliban and decided to move outside into the shade to cool off in the hilltop breeze. Noon- it is my turn for a 2-hour radio watch. 25 minutes before I was supposed to get relieved, I heard the sound of a vehicle right behind me. I turn to the left to see two men in a blue truck driving slowly beside us, and my heart jumps to my throat. I yell for 1st SGT, and he gets on the net and calls it into the Marines ahead of us. They rushed out to intercept, but by the time they got to the hill, the vehicle had already gone too far out of sight. 1st SGT ventures up the hill behind us to investigate how the truck got past the Marines on watch behind us and can clearly see them in plain view. He comes back down and says, "Someone must not have been paying attention back there." I am pissed off thinking that could have been the death of us if those guys were our enemy. We relocated to a hilltop that overlooked all of the land and sat there for hours. I am now 2/3 way through my Tom Clancy book, and thank God for the continuous cool breeze. 17h30 rolls around, and we head back to coil up with the rest of HQ and Blue platoon for the night. Chris, our Afghan interpreter, comes to me for headache medications, and I lay my head to rest while listening to fellow Marines talk about what they want to do and who they want to see when they get home. As I fell asleep, I couldn't help but think about my special someone at the time waiting for me when I got back home. In the middle of my slumber, I was woken up by Cruickshank to repair his face injury. He has a busted upper lip. I did my thing and placed tape over the bandage for him. I even gave him a pencil-thin curly mustache drawing for happy enjoyment, and then I went back to sleep.

September 7 – 05h00 reveille. Bravo HQ strike force heads out to recon an area to cross the river. We drive for hours through mountainous terrain passing steep cliffs searching for an open route. The fear of tipping and rolling in a giant LAV was exciting and nerve-wracking. Imagine rolling down a steep cliff uncontrollably in a giant metal box with all sorts of crush hazards and no seat belt. We come across an area next to a village hut full of women and children. We decide to travel south and look for an easier route. After about an hour, we find a safe route and cross the river. Upon getting across, we stop at various small villages and tactically question the elders of each village. We also came across several fields the size of 3-4 football fields full of marijuana. We all have never seen so much pot in all of our lives combined. We decided to pull off between the fork in the river and relax for a while, throwing the bumper toy for Molly and watching her jump in the river to get it and skipping stones. Chris, our Terp (interpreter), decides to strip down to his boxers and go for a swim. The rest of us got the green light to join in on some river swimming fun. It was a relief to be in the water. It was a good time for everyone, and it definitely made my week and boosted my morale for being out here. Several hours pass, and the sun sets. We drive through the water to cross the river and through the thick vegetation up the mountain to the top and coil up for the night. The view overlooks the entire river valley. It was beautiful. We set up a watch schedule and bed down for the night. I got notified that I have first watch. I sit in the turret of Black-4 overlooking the river valley from the edge of the cliff. This watch was interesting; I was sitting there playing with a multimillion-dollar piece of equipment using thermal image scope sights, watching our route, making sure no one was performing suspicious activities. I could clearly see many animals, including a flock of camels migrating from the shrubs towards the riverbed for a drink. It was the coolest 1.5 hours of watch I've stood thus far.

September 8 – 05h00 reveille. The mornings are getting colder. We pack up right away and travel back over the sand hills to link up with HQ Bravo Section to resupply and do vehicle maintenance. Hernandez brings Molly to me as I am helping Orta change a flat tire and says her ears are bleeding due to them cracking from dry skin. I take out my medical pouch, clean her ears, and put bacitracin and tape up the tips of her ears with medical tape. Poor girl now has white-tipped ears. It looks kind of funny but saddening at the same time. I felt good, like an animal veterinarian, while treating her. We then packed up and headed back out over the sand hills down to the river, crossed it, and linked up with Red Platoon to relieve their position along with the riverbed so they could resupply. I ran out of the food in cans that Mom and Dad sent, so I ate an MRE and relaxed in the pig, petting Molly and making sure her ears still had tape on them. Villagers herding sheep and goats approach us, and Chris is directed to shout to them in the Farsi language to stay away. We continue to hold our position on the riverbank. I work out with Sgt Abelar and do it with my shirt off so I can catch a little tan. 1st SGT gives us the okay to take a drip in the river. Orta, Hernandez, Boslee, Molly, and I strip down to our shorts and go for a swim. The water was murky brown, but it was refreshing. The floor of the river was coated in soft mud, and it felt like literal shit in between my toes, but I didn't care. It felt so good to be floating in water for a few minutes again. We all hopped out and walked 20 feet back to the pigs soaking wet. By the time we reach Black-8, we are completely bone dry; that was how hot and dry the environment was there. Wagoner gets finished digging the hole for the burn pit and sees us putting our clothes back on, and shouts, "You assholes didn't wait for me!" I laugh and say, come on, Tommy, let's go. I will provide security for you. I noticed that Molly's bandages had come off of her ears, so I cleaned them again and bandaged them back up like a good vet/doctor. Around noon time an old man approaches our position, and Chris yells at him to lift his man dress so we can see he does not have any weapons hidden. The man is not compliant and does not want to cooperate. I grab my M4 out of the back of the vehicle and point it at him. The notion of looking at a barrel pointing at you seems to be a universal language. Now he understands, but apparently, the people on this side of the river speak a different dialect of Farsi, so Chris sometimes has a hard time with the villagers when they talk. We continue to relax and switch out security posts while the noon sun beats down upon us. The old man continues walking away alongside the river in the far distance. I break out my last bag of noodles and put them in a beverage bag and fill it with water and place it on the hood of the vehicle in the direct sunlight to cook. They will be done in a couple of minutes, and I will have my lunch. While I was waiting for my noodles to cook, a couple of us decided that we wanted to cool off again in the river. This time I convinced SGT Abelar to join Hernandez, Capt. Haschole, Orta, Molly, and I for a swim. We cooled off and decided it's time for some soaking wet skivvy shorts poses for pics. We all hike up our shorts and fold them in at our hips to make it look like we are wearing speedos. Me being crazy like I usually am, I decided to hike up my shorts in the crack of my ass to make it look like I have a thong on. We take some pictures and goof around and head back to the pigs on the hot rocks stepping lightly and stumbling on the sharp edges trying not to step in the massive amounts of camel shit scattered everywhere along the riverbank. I get back, get dressed back in my FROG (Flame resistant operational gear) uniform, and eat my lunch that has been cooking in the sun. I sit in the shade, enjoying the cool breeze wondering how the rest of the company is doing. They probably hate life sitting in the hot sun and working on getting other vehicles unstuck from the sand dunes with much lower morale than us. Even though we were out here having fun with higher morale, I still tend to miss being back home and seeing loved ones. 13h30, I sat on my chair and read my book for a few hours in the pig and tried to take a nap. I had trouble trying to fall asleep due to the heat flaring from my sunburnt back. My Marines are relaxing as well, still trying to stay out of the sun. The camel and sheep shit attracted flies that continued to pester me, landing and crawling all over me as I tried to relax. We packed up at 17h30, jumped back in the rolling pig, and moved to the HQ coil. On the way, we pick up several Marines from Blue platoon due to their vehicles breaking down. We get to the coil and talk to Dev, Mantis, Hermon, and relax as the sun goes down. I have radio watch on the C2 (Black-5) from 20h00-21h00, then slip off to bed.

September 9 – 05h00 reveille. Everyone packs up and prepares to head to the west to screen line and interdict vehicles. We take 2 other Marines from Blue platoon to work with us. 07h00, we get radio checks, load up, and head out into the desert. We drive until 10h30 stopping 2 trucks and searching them on the way. We finally set a spot in the middle of a roadway. I set up a piss trench and dig a fire pit for trash. Out of boredom, Dunaj, the driver for Black-6, comes and helps me, and we decide to decorate the holes with rocks. Several minutes later, we spot 31 camels [yes, I counted] traveling in a straight line towards one direction, followed by an old man and a young child. We stop them for questioning then and letting them continue once we see they pose no threat to us. Noon rolls around, and it is my turn for a 2-hour observation point post. I sit there in the direct sunlight, scanning with binoculars. 14h00 Hernandez relieves me from post, and I find a shaded spot next to the vehicle and try to relax as I swat flies away from my face. Molly comes and lays down next to me for a while. 16h00, we decide to break our boredom and place bets on several Marines in a 400-meter lunge race. I was the best money holder/referee/safety Corpsman. After the fun was over, the CO gave us a brief on tomorrow's operations we are doing in Deeshew. I sit to play chess with Hernandez then I read another chapter of my book. 18h30, I decided to continue reading and finish the book.

Chills ran down my spine as I finished off the last few paragraphs. Upon finishing the last sentence, I looked up to the endless desert to watch the sun set. I can't remember the last time I had the chance to sit down and read a good long book, but it felt great. I sit there watching the sun disappear behind the horizon and am already starting to crave another good book. 17h00 rolls around, and I lay my head to rest but as soon as I do, I overhear SGT Abelar shout to us that he spots 6-9 lights on the horizon through his NVG's (night vision goggles). We all grab our NVG's to observe, and by the time I get mine and observe, there are now 15 lights, and we continue to observe at first thinking they are vehicles, but then figure it is probably fires for all the villages to celebrate their last religious day of Ramadan. 17h35 I have stand down, and I decided to get some sleep. Just as I lay my head on the medical pouch I use as a pillow, I hear 1st SGT yell, "WTF!" He then said, "A coyote just walked right up to me, and I went to pet it thinking it was Molly, and it ran away as I put my hand out to it." Everyone started laughing, and shortly after, I fell asleep.

September 10 – Midnight I was woken up to stand an hour of watch on Black-4 which went by fairly quickly. Hernandez relieved me, and I started walking back to the pig when I noticed Molly was wandering around, so I told her to get in the pig and as I lay to rest she jumped on me to cuddle. 05h00 reveille. We pack up camp and head back to HQ coil to resupply as we wait on the clifftop for Red Platoon to clear a route so we can head to a suspected HME (Home Made Explosive) lab in one of the villages. I asked Devin how he was feeling today, and he said the medications I gave him for his congestion worked well. 09h00, we spot a motorcyclist heading west, so Black-8 gets geared up and rushes to intercept. Carbo, another Marine, climbs in the back with me to assist, and we catch up with him and search him. He knows the routine, so it was fairly quick, and we head back to the coil. I ask the XO if he has any Tom Clancy books, and he says he does, but he must search for it, which means I am going to have to keep reminding him to look for it. We receive a BFT message from Highlander Actual (top chain of command) that a large storm is rolling in. We also received radio traffic that 2 Red Platoon vehicles have gotten stuck in the sand during their process of clearing a route for us. We decide it is time to move out to recover them. We take Black 10 with its front mine roller attachment, truck 2, and Black-6 with us to head to the river. Black-8 and Black-10 hold security on the north side, while the EOD vehicle holds security on the south side. Black-6 and truck 2 push further to recover Red-2. While we are stopped, 2 dogs approach us with curiosity, and they look like hyenas. The dogs lose interest and walk away. 11h30 I proceed to melt in the sun while awaiting the next orders from the CO. 14h00 Black-7 and truck 2 cross the river, park next to us, and drop off the packs and gear of 4 Marines from one of the broken down Blue platoon vehicles. Then we cross the river to link up with EOD and hold until we receive word to link up and hold with the rest of Blue platoon. I listen to radio traffic once we park and overhear Blue platoon finding a mortar round or 60mm like shell in the ground. 15h30 Blue-1 comes to our position and escorts us into the villages. We stop at an OP where Webster, Head, Woodrow, and Carter are sitting in a hole in the ground, keeping an over watch on our route, making sure no one decides to dig and plant IEDs. No one has any medical complaints, but they are all thirsty and hungry, so I give them some food and water from my personal stash and set up shade for them in the pig. Just as the Marines and I set up a watch schedule and sit down for a bite to eat. A lone camel walked right up to us with curiosity and then swiftly walked away. I observe my surroundings; we are located smack dab in the middle of several villages. There were mud-made buildings surrounding us from all angles. Traffic was constant here. Every 5 minutes or so a small car or travelers on foot will pass and the Marines will stop and search them. 16h00; blue platoon scouts report Molly sniffed out a possible IED. Black-8 is tasked out to escort EOD to it. We drive for about an hour, passing several large villages, including the village with that Huge Castle from the last operation. I notice we are in a familiar place due to its massive size and shape. We link up with Black-6 holding a perimeter around the possible IED (PIED) site. There in the distance over the dirt hill, there are 2 sticks stuck in the ground pointing straight up, some wire in a plastic bottle, and disturbed dirt on the site. EOD personnel jumped out of the Mine Resistant Ambush Protected vehicle (MRAP) they were maneuvering in. They travel in and start investigating the site with minesweepers. I sit and hold rear security and await the order to lay low if they decide to detonate it or if it unintentionally detonates on them. 16h30, the investigation is complete. It was not an IED, and it turns out it was just some trash conveniently near 2 random sticks that the Afghanis used to mark routes for travel. We load up and head back to the river villages. 18h40; we arrive at our previous OP spot to link up with the Marines on route post, and we prep for the night. 23h00 to midnight; it was my turn for post. I grab my gear and M4 and jump on the top of the EOD MRAP Turret. I scanned 360 with my night vision goggles as I listened to the sound of dogs barking to each other from each surrounding village and the sounds of my Marines snoring while sleeping in the dirt on the ground below me. An hour passed by, the wind chill caused me to shiver during watch, so I grabbed my poncho and wrapped myself up in it, and crawled back to bed after my relief came. I lie on my back and fall asleep watching a small meteor shower in the night sky.

September 11 – 05h00 reveille. September 11 – 05h00 reveille. We got the order that we are going to make movement back to Camp Payne today, so we packed up and prepared for movement. Blue scouts with us get the order to sweep into a village for key leader engagement and tactical questioning at 06h30. They began their patrol out to one of the villages next to us. 07h15; a vehicle comes in the direction towards our position. It was a black truck with an old man and a young boy inside. It was the cleanest small little pickup truck I had seen since arriving here in Afghanistan. The truck is searched, and Wise, the interpreter with Blue platoon, comes over to translate the words from the old man. He claims to be the elder of one of the villages and owns the ferry boat that crosses the river. He gives us good information on the Taliban activities in the area, and we send him on his way. 08h30, we hang out back at the vehicles as the sun rises higher in the sky over the desert. 10h30 Black- 4 and Black- 6 meet up with us, and we begin our push back to the HQ coil to refuel before our convoy back to Payne. On the way down to the river's crossing point, Black-8 gets stuck in the soft moon dust on the route. Orta tries to maneuver out, but since we have 6 other marines, including myself, all of our packs and gear, the vehicle is much heavier than usual, and we sink further into the ground. There was nothing we could do at this point except one of the many things Marines do best, we dig. For hours we dig in the hot sun around the vehicle with moon dust everywhere while we wait for the 7-ton truck to come to pull us out.11h45 the 7-ton arrives, and we rig it for tow. Just as we were all ready to tug, Black- 8's fuel pump goes out, so it's a pain in the ass to pull because the wheels lock up unless the vehicle is on. Eventually, we get some mechanics to fix the problem, and we head back to Camp Payne. The drive seemed like it took forever because we drove 20kph the entire way. We reached Payne just after the chow hall had closed, but Blue platoon grabbed hot trays for the rest of the company. It felt good to get a chance to shower. Later that night, I took a moment of silence to myself, thinking about all of our fallen brothers in the 9-11 event 9 years ago.

September 12 – 07h00 reveille. Down the ramp for vehicle maintenance. That is pretty much what we are going to be doing for our time here at Payne. Computer, phone center, chow hall, mattresses to sleep on, air conditioning, and of course putting up with bullshit from the rest of the battalion that is here. The sandstorm rolled in, and visibility went down a great deal outside. Funny side note, I shaved Devin's ass on a dare out of boredom later that evening, haha. When we got bored, we did crazy things to keep morale up. I received a package from Mom and Dad. His homemade venison jerky and mom's zucchini bread were superb. I also received 3 packages from Joanna with canned food for the field.

September 13 – 07h00 reveille. Back down the ramp for more vehicle maintenance. Also, a full EDL serialized gear inspection. I can feel the boredom kicking in at this place already, but at least we have all these nice things here for us, so I can't complain. We got word passed today that we were soon going on a 2-week operation. We are going to be replacing Charlie Company in a coil outside of South Station across the river. That doesn't sound exciting at all.

September 14 – Nothing too exciting to talk about today except that Doc Mantis earned his FMF pin today. We packed our belongings for operation, and that was about it.

September 15 – 06h00 reveille. We gather the remainder of our belongings and head to the ramp for radio checks. 08h00, we head to South Station just across the river from Payne. We set up our living spaces in the tents and set up a work schedule. This place is a sandbox. There are members from the Afghan Army here standing security posts and living with us in separate tents. Living conditions are fairly good except for not having air conditioning, running water, chow hall, computer center, phones, or mattresses to sleep on. There is a stray but friendly Afghani dog here named Nixon who just lays around all day. The temperature is dropping, and it is not as ridiculously hot as it used to be during the day unless you stay in the direct sunlight and the mornings are cold. The only thing we are doing while being here is standing post and acting as QRF (Quick Reaction Force) in support of the rest of the company patrolling outside this place while Charlie Company does reset training at Payne. Everyone is starting to talk about going home and what they plan to do upon getting back because, excitingly, we only have a little over one month left in this country. I myself have a couple of things in mind that I am looking forward to when I get back. I sat around pretty much the entire day doing nothing except watching movies on the laptop, listening to music in the tent and masturbating a whole lot to pass the time. Every now and then, I will see a rat run across the floor scavenging for food. It is a good thing Hernandez let me borrow his laptop, or else I would be losing my mind in this place. He had a lot of great porn on it.

September 16 – 07h00 reveille. Last night I had a dream that I could fly, which was awesome. I received word that I will be needed on a mounted patrol outside the FOB so we can drop one of Red Platoons, formerly down but now fixed vehicles. Black-8 is still being worked on, so I ride in Black-7 with MSGT. The ride is boring, and one of the MRAP (Mine Resistant Ambush Protected) vehicles breaks down, so we end up sitting around for a few hours on a dirt road providing security until the mechanics can fix it. Noon, several vehicles traveling on our route get stopped and searched by Marines. I was called to assist in the interdiction process on the 4 blue trucks that were traveling together when they approached. We stopped them for questioning. A total of 8 men, one woman, and one child were traveling fuel salesmen with empty jugs, bags of flour, and food. We searched them and then tactically questioned everyone and sent them on their way. We met up with the rest of the platoons at each town they were covering, passing several large villages filled with trees and vegetation on the way. Most of the people in these sectors are just simple farmers. We finished resupplying the company and headed back to South Station. I unpacked then went to the makeshift outdoor gym that was established there. Afterward, I helped Cesar Mas and Orta change out the fuel pumps on Black-8 in the maintenance bay then I hit the rack.

September 17 – 08h00 reveille. The Non-commissioned Officers (NCOs) and I moved into our own separate tent from the boots, and that's pretty much where I stayed for the rest of the morning watching the movie "Forest Gump" until 13h00. I went back to the maintenance bay and finished organizing the back of Black-8. 17h00, I finally got to meet HM3 Donahue. Nice guy from California. Chief granted him to come and help out with the company. I could always use an extra set of Corpsman hands. I get word that I am going out to sit in a coil outside of the base with several Marines to monitor traffic in a close area for the night. The sandstorm has been pounding us all night, and we are going out to see if any Taliban are trying to use it to their advantage because we usually don't operate during red air. 18h00, we pack up and move out to the desert. We stopped at the previous IED site between South Station and Camp Payne and set up a small coil. The sun has set, and the CO gives us a short brief saying this is a historic IED site; if you see them digging, kill them. I stood post from 23h00-00h30, and I was freezing my ass off. I get relieved 15 minutes late and try to fall asleep snoopy style on top of Black-7 wrapped up in my poncho liner wearing a light desert jacket, but the wind was blowing, and it was way too cold. I ended up staying awake the entire morning with no sleep. Reveille was sounded at 05h30, and we packed up and headed back to South Station. Upon arriving at my tent, I found my iPod lying there, so I picked it up and found that it was broken internally. I was pissed because I wouldn't have any more music to listen to for the remainder of this deployment. I am too tired to care right now, though, so I lay my head to rest. 11h15 I wake up to the feeling of rats crawling over me and SGT Abelar yelling because rats had chewed through his bag and ate holes in his protein powder packets. This place is disgusting. I go outside to piss and see Nixon chewing a dead fox skull near the piss trench. I go back to my rack and get some more rest and see rat feces surrounding my entire rack and all over the floor. I clean the area up while gagging and lay back down to sleep. Nope, I can't sleep. It is hotter in these tents than it is outside, and the AC unit is a loud squeaky good-for-nothing piece-of-shit. So, I went to the maintenance bay to keep myself busy and crank wrenches for a couple of hours because our 14h00 mission got canceled. Everyone continued to relax in the tents watching laptop movies and trying to keep the generator working. Word got passed for us to stay on our toes because Castle (one of our other forward operating bases that Alpha Company occupies) just got rocket attacked by members of the Taliban. I am so bored of this place. I sit here and wonder if we would be ready if we got attacked with rockets here. I also still can't believe we all made it out of Derzay, Operation Red Dawn, without a KIA from Bravo Company. Our marines were either that good or just that they simply had the luck of the devil himself.

September 18 – 03h00 I awoke to the sounds of mortar rounds exploding. I ran outside my tent to see what was going on when I saw a bright flash of light falling from the sky. Several more rounds went off, and I had then recognized the sounds to be that of friendly mortar Illumination rounds. Apparently, one of the posts on guard thought he saw people sneaking up on him. I went back to bed and awoke at 09h30. I had asked what the commotion was all about earlier, and it was for that reason. I got something to eat from the chow tent, and there is where I sat all day watching movies on the laptop. CPL King and I noticed Nixon limping, not using his front right paw because he stepped in our mousetrap yesterday. I offered to wrap it up but didn't feel like getting bitten by a dirty Afghan dog that uses rotten varmints as chew toys. I ended up just going to sleep.

September 19 – HQ Bravo team packs up the pigs with supplies for Blue platoon at 07h00, and we push out their position at 0800, stopping at Payne across the river along the way. We return in the afternoon, and I sit around doing nothing all day after that.

September 20 – 06h00 reveille. We mount up to resupply the platoons and drop off BATTs and HIIDE (Biometric data collection systems) to them. I met up with Tien and we shared a couple of minutes of conversation while overlooking the towns that White Platoon was doing patrols in. After meeting with all of the platoons and seeing my junior Corpsmen, I noticed a change in everyone's attitudes out here. Morale amongst the company has plummeted. I guess everyone just wants to get this deployment over with already. We returned to South Station later in the afternoon, and I didn't do anything significant except go to the sorry excuse for a gym that was here and use the satellite phone to call home.

September 21 – I awoke at 09h30 and saw 2 letters from a girl I was talking to from California on my bunk. I was pleased to finally receive mail and went to chow in a good mood because of it. I literally sat around all day today doing nothing except watching movies on the laptop. HQ Alpha section punched outside the wire at 15h00 to conduct an overnight operation at Blue platoons' position. I continued to enjoy peace and quiet in my tent while sweating my balls off for the rest of the night. 22h45 SSGT Stine rushes into our tent and informs us that we are conducting an HE (highly explosive) mortar attack on a Taliban member plating IEDs next to Payne. Minutes later, several loud booms go off and silence then the sound of explosions in the far distance across the river. The excitement was soon over, and I crawled into bed.

September 22 – I slept in until 11h00 this morning, and I don't feel bad about it at all because there is nothing to do here. One of our mechanics came into the tent around noon and showed me that he had a 1.5-inch laceration on his left calf muscle. I patched it up and informed 1st SGT that he needed to go back to Payne for stitches. I sat back on my ass and continued to do the usual, watching movies and listening to explosions go off in the distance.

September 23 – 07h00 reveille. Everyone cleaned up their tents and tightened up the place, and made extra sure everyone else had clean shaves and was wearing their uniforms properly. The battalion commander was coming here from Payne to conduct a meeting bringing with him the battalion Sgt Major. After sitting around for several hours trying to stay out of sight and out of mind from the battalion leaders, Bravo section HQ loads up at noon to collect mail from Payne to pass out to the platoons after the Bn CO and Sgt Major depart. We sit on the pigs for over an hour, waiting for the vehicle commanders to show up, but they never do. Someone with a radio finally asks if we're still going, and they reply to stand down. It's this kind of bullshit that makes me not want to re-enlist. No one knows what we are doing or waiting for. We head back to the tents after locking the pigs back up. I notice several Marines digging a burn pit and making chairs out of sandbags for a campfire site. It is now 18h00, and I feel like another day has been wasted away where I did nothing.

September 24 – I woke up at 11h45, happy that half the day was gone already. I felt sluggish though, so I went to the gym twice today. I sat around watching movies all afternoon, and that's all I have to talk about other than getting a third letter from a girl back home today. I have never been so bored, but it is nice to get paid to relax all day. We also set up mouse traps, so I am waiting to hear a loud snap in the middle of the night tonight. 23h45 snap! Yeah, we caught several mice throughout the night. Not very exciting, but at least there will be less mouse pellets and fewer holes in all of our food from the mice eating it all.

September 25 – 07h00 reveille. We load up with boxes of mail already on the pigs for the platoons. MSGT comes over, telling us to help him catch Nixon. After a few minutes of searching, MSGT finds him and puts a leash around him, pulling him to the MATV as he drags his hind legs resisting getting on the vehicle. We finally get him on and close the door. I ask what we are going to do with him, and MSGT says we have to get rid of him. I load up, and 1st SGT notifies me that I need to tell one of our Marines from Red Platoon that his mother had been diagnosed with breast cancer. (Thereafter, I would periodically follow-up with this Marines to ensure he was mentally stable about the terrible news until it was time to send him home to be with his mother). Once that was completed, we drove off at 08h15 crossing the river and passing through Payne on our way to drop off mail to Blue platoon. We stopped in the middle of the desert and unloaded Nixon, and closed the vehicle doors. I watched him stand there, looking confused as we drove away. He began running in our tracks, following our trail hobbling along with his injured leg, trying to keep up, but our vehicles drove away too fast for him to keep up. I felt sad but wasn't worried because there are villages nearby that he could make a new home with. 10 minutes later, we reach Blue platoon's position and start unloading their mail. About an hour goes by, and we spot Nixon still following our tracks. He eventually sees our position and gallops over to the MATV, and waits outside the door of the vehicle. I could hear explosions and gunfire in the distance, thinking something was stirring up in one of the villages, and then soon realized that it was probably one of the Marine firing ranges at Payne. We sat around for a couple of hours, waiting to move to the next platoon. We traveled to Red Platoon, stopping at Payne, and had time to grab some good chow hall food. Eventually, we made it to Red's position and dropped off mail. We then traveled to White's position, where I was greeted by my good buddy, Tien. I handed him his mail, and we caught up on current events about ourselves and talked about home for a few minutes. Then I went over to Doc Raymans' vehicle and handed him his mail, and asked how everything was going. He was excited to show me the chicken his Marines were temporarily keeping as a pet before they ate it. I then went back over to Tien to say goodbye for a couple of days, and we packed up and rolled back to South Station. We arrived at 1900, and I geared down and prepped for a good night's rest. My morale had been boosted due to seeing my other friends and actually doing something productive. Wagoner informed me that Alpha section of HQ was going on a dismounted patrol to a nearby town traveling on route blood stripe, a well-known heavily IED sabotaged roadway. I personally thought it was a bad idea to be doing this at night, but no one was asking my opinion, and it didn't matter anyway. I then found something to eat then laid to rest. As soon as I fell into a deep sleep, I was awoken to do a counter-terrorism attack drill. Yeah, I was pissed and tired, and of course, cursing the situation while going to the vehicle and getting all of my gear on to stand on a guard post. Loui O was there with me on post, along with a member from the Afghan Army. This guy looked as if he was 80 years old, and he didn't speak a word of English, but we made the best of the situation. We were showing him how to use night vision goggles, and he was teaching us his language. I got this idea that we could teach him our language, but we were saying stupid, vulgar things instead of for him to repeat for our entertainment. We heard the call to stand down, and we went back to our tents to rest. It turns out, this day was quite terrible because I had to comfort a Marine after giving sad news about his mother having cancer, we had to abandon a friendly dog with an injured leg in the middle of the desert, and rest was disrupted for bullshit training after a long day of operations.

September 26 – I woke up at 09h30 and received a package from the Beairsto family with snacks and crossword puzzles and magazines to read. I missed them so much. They have been so full of love and laughter from the moment I have known them. I went to the gym then watched movies until 19h00. We had a counter-terrorist attack class/brief in the gym area, and I went to bed shortly after.

September 27 – 06h30 reveille. We went to the pigs. 08h00 traveled to Blue platoon to pick up 2 vehicles to do maintenance on. We stopped at Payne to talk to Chief about flu shots. 11h30 traveled to Red Platoon to pick up a member from EOD, Hernandez, and Molly. 12h15 drove to White Platoon to conduct a patrol in one of the towns. We received intel that there are 7 IEDs placed in a specific area in one of the towns here, so the Marines are conducting a patrol to check it out. After several hours of waiting for the patrol to get finished, the Marines returned to the coil with no finds and nothing unusual to report. Hernandez, Molly, EOD, and a Terp jump in my pig, and we travel to red position to drop Hernandez and Molly off and take the other 2 back to South Station. As soon as we unload Molly and Hernandez, White Platoon reports contact at 18h00. We scramble back to White's position because they reported being rocket attacked. 18h15 reach the outskirts of White's coil where members of White Platoon are investigating the explosion impact sight and where it came from. We park less than a football field away in a small coil with White Platoon. The sun has set, and visibility goes down as EOD further investigates and also comes up dry. We're planning on staying here for the night, expecting another attack. After getting set up in White Platoons' coil, I make my way over to Tien and get the bum scoop on the event. He told me there were 2 rockets that exploded on both sides of their coil. No one was injured due to the Taliban's aim, but they were close considering they were fired from about a km away. We set security and made ready for a secondary attack at 19h00. 20h00, we stand down and I drop my gear and head over to Tien's pig, where we converse under the camo netting until 21h00. Sgt Hernandez came over and notified me I had radio watch on Black-4 for an hour. 22h00 Orta comes to relieve me from post, and I head back to Tien's pig to keep him company to help his 3-hour post go by faster.

September 28 – 05h00 reveille. We leave White's position to get resupplied at Payne. 08h00, we arrive at Payne, and I head to the BAS to get resupplied on medical gear. HM2 Delacruz hints to me that Chief is putting me up for sailor of the quarter after not telling me why he needed pictures of me standing at attention. We resupply each platoon throughout the day on chow, water, and fuel. I resupply all of my corpsmen on the gear they requested, and we head back to South Station and head to our tents at 17h00. While I was ground guiding my vehicle to its parking space, Wagoner asked me if he could borrow my Gerber multi-tool, and I tossed it to him, but he wasn't paying attention, so instead of catching it, it hit him right in the face cutting his forehead. I patched him up and headed to the tent. Several minutes later, we get called out to escort EOD back to White Platoons' position because they uncovered an IED. 18h00, we load up and head out. 18h30 we arrive, and EOD punches out to investigate the site. I head over to Tien's pig to chat with him about my day and Wagoner's incessant stupidity. An hour went by, and we heard and felt a loud explosion in the village we were overlooking. It startled us, and we overhear it was a controlled detonation that EOD prepped to demolish the pressure plate IED they found. Several minutes later, another one that they set to eliminate the remaining parts of the IED goes off. I head back to Black-8 and wait for EOD to return to the coil. 21h45 EOD returns, and we head back to South Station, arriving at 22h30, where I head to my tent and rest for the night.

September 29 – I awake at 09h00 feeling like I still need more rest to catch up on. I decide to hit the gym for an hour. Noon; 1st SGT comes to me after I get done working out and tells me to load up on the pig. We are heading out to pick up a detained member of the Taliban. At 12h30 we reach whites position and prep the back of the vehicle to transport the detainee. Zacky comes up to Orta and gives us the low down on the 2 men they captured. He said they were on patrol and saw these 2 men acting suspicious, so they chased them down and searched them. The one-man had letters from the Taliban and was in association with terrorist activities. The other man had nothing on him. We hung around in the shade swatting at flies for a few hours, waiting for word. The detainee gets totally 'blacked out' with painted goggles and earmuffs before being loaded up in my vehicle along with 2 Marines guarding him. We head to Payne and bring the Taliban member to the BAS for a physical then to see the judge. The good thing about being here for several hours was having a nice meal at the chow hall. After eating, I head back to the pig at 17h30, where I wait for the others so we can head back to South Station. About an hour of waiting on the pigs at the ramp goes by before everyone shows up. We get radio checks and push back across the river. We arrive at South Station, and I grab an MRE, eat dinner, and try to fall asleep but can't. I get up and keep Devin company on his post, conversing with him until 03h00, then I go to sleep.

September 30 – I awake at 09h00 due to the intense heat in the tent. I grab something to eat and then watch movies with Orta on his laptop for a while. Noon rolls around, and a few Marines spot a venomous snake slithering outside our tent. We chase the snake down and kill it so it doesn't harm anyone who might walk by and get bitten, then we throw it over the berm. I get informed that I have a phone call from the BAS at Payne, so I go to the COC and talk with HM2 Delacruz about shots and Marines who need follow-up appointments. He also tells me I need to email a letter of bibliography from when I graduated high school until now and all things significant about what I have done for the Sailor of the quarter package they are trying to submit me for. I send the email within the hour and come up with a plan with 1st SGT and MSGT about getting everyone in the field flu vaccinated. I then head to my tent to grab a bite to eat for lunch. It is very hot and strangely humid today. Everyone is complaining about the heat, and I am sweating my balls off while trying to eat lunch. Several hours go by of sitting around. HQ Alpha section punches out for an overnight operation at White's position. I hit the gym after spending an hour on the satellite phone with Dad letting him know I was alright and telling him to stop letting the news get him so worried. After the gym, I grabbed another MRE and sat down to watch movies on my rack. Just as I do, Cruikshank comes in and says I need to look at Boslee's foot. I grab my medical pouch and head over there to find him sitting on the deck of his tent, looking at his left foot. He had stepped on a nail sticking out of a piece of plywood. I flushed the hole in his foot out with alcohol and told him to wear his boots from now whenever he walks around. I went back to my rack and found myself feeling sleepy, so I just passed out feeling glad that September was over.

Chapter 11: A Month That Would Wake Anyone Up

October 1 – I awoke. It was 10h00, and I took a shower in the makeshift shower pit they had built here. It was placed across the way in view of the Afghan Army unit we were sharing this base with. It felt good to clean up with water and soap for a change. Baby wipes just weren't cutting it anymore. At 11h00, it got miserably hot again today, and I have been doing nothing except wishing I were home. I have been soaking wet with sweat all day, hating life. Word on the street is that we will be staying in this sandbox for the rest of deployment, but things are always subject to change in the Corps. I placed a can of chunky soup on the berm in the sun today, and within 10 minutes, it was hot to the touch, and I had to wait for it to cool off before eating it. Several hours went by as I sat around doing nothing. I found some colored markers and pens and decided to get creative and color some crafts. There were no crayons, so the Marines reading this can lay off that joke with disappointment. The sun goes down, and it cools down enough for me to comfortably lay my head to rest.

October 2 – 07h00 reveille. We receive word that we are escorting the battalion commander out to White's position so he can check out the area. Also, SGT West is getting promoted to SSGT today, so the CO is probably here for that as well. I met up with Tien to see how he and his Marines were doing. LCPL Morgan has been throwing up all day, so I give him anti-nausea medications and tell him to rest. It is noon and miserably hot again. I am sitting in the vehicle because it's the only thing providing shade right now. 14h00, we push to Red's position. There are bulldozers and other construction vehicles here building a VCP (vehicle control point). I put noodles on top of the vehicle an hour ago, so I stayed in the vehicle and ate after throwing out reds' mail to the Marines. 14h00, we pack up and head to Payne to drop off the CO for his meeting about our possible upcoming mission in Bahram Cha. We load up shortly after and head to Blue's position to drop off their platoon commander. 17h00, we head to Payne so 1st SGT can do paperwork. I head up to BAS to drop off the shot roster for flu vaccines to HM2 Delacruz, and he gives me new medical bags to pass out to my Corpsmen. We eat hot tray meals on the ramp and sit around for hours waiting for the XO to get done doing his business. 21h30, we head back to South Station, and once we get there, we hear about a Marine in Charlie Company that was on post there had negligently discharged (ND) his weapon; luckily no one was injured. As soon as I got to my tent, I went to sleep.

October 3 – 09h00 I awake to the rise of temperature and the sounds of my Marines yelling at each other in an argument or something. I hygiene and head to the COC to call BAS to discuss the plan for giving shots to the company. 11h00, we load up on the pigs and head out to Payne. 13h00, we head to Whites' position after picking up our recent detainee. The judge found him innocent, so we had to let him go. It was frustrating to everyone that we did all that work just to let a bad guy go. We arrive at 13h30 and then head back to Payne. HQ Alpha section heads to Payne to stay there for 10 days to prep the company on vehicle maintenance and medical training for our last big operation. This Bahram cha operation is going to be pretty kinetic and dangerous, so our main focus from here on will be in preparation for an epic battle. Feelings of excitement and nervousness rush through my body, but as things are always subject to change, I am in disbelief of anything at this point until it actually happens.

October 4 – Woke up at 09h00. SGT Hernandez gave me a haircut, and I went to the gym. Noon, I eat a can of tuna and sit around doing nothing until 19h30. 19h30, we prep and go over the plan for an attack reaction drill. SGT Abelar and I run around to different spots in this sandbox, picking Marines to become casualties as notional rounds drop on us. It lasted about a half-hour, and the medical portion of the drill went smoothly. It is now 21h00 and still too hot to get sleep, but I try anyway. Sometimes I would lay there all night trying to drown out the pain and suffering of the heat. Other nights I would do my best to think of positive things that may come in the future in order to achieve somewhat pleasant dreams before I would fall asleep.

October 5 – 06h00 reveille. We load up on the pigs, and at 07h00, we push out to Payne, and I head up to the BAS to gather the flu shots. 09h30, I have 6 hours to administer these shots to as many Marines as I can before the heat of the day spoils the vaccinations I have in this icebox. We push to White's position and I administer shots; an hour later, we push to Red's position, then to South Station to finish the rest of the company. It felt good stabbing certain people with needles as if I were relieving stress. 14h00, we push back to Payne, and I drop off the shot forms to HM1 and debrief him. As soon as we start rolling back to South Station, White calls up an IED 9-line, and we push out to their position again. 16h45, we reach the outskirts of White's coil and push EOD out to the sites of 3 IEDs. The sun is setting behind the mountains, and prayer is heard on the loudspeakers in the towns. I watch Orta and Chris, our Terp 'ground fight' in the sand, and await the news from what EOD is doing and their progress on the IEDs. As we waited, I walked over to Tien's pig, and we both lay with our backs in the sand talking about the excitement of going home and looking at the stars thinking of people back home. I saw 5 shooting stars within 2 hours. If there was one good thing about this place, it was the stars and the sunsets. 19h30

EOD calls in the shot window for demolishing the IEDs. 20h00; 3 loud explosions go off in the distance, and I head back to Black-8 to await the movement back to South Station. I lay on my stretcher and await while watching the stars. I fall asleep while waiting and am woken up at 01h00.

October 6 – 01h00 I was awoken to stand 30 minutes of turret watch on Black- 4. It seems like the higher-ups made the decision to spend the night here. I guess I was so tired of battling with the higher-up Marines about the whole gun watch ordeal that I decided to say fuck it, suck it up, and sit on the gun for the sake of helping my other Marines get some rest. 01h30 I get relieved and walk about to Black-8 shivering cold and try to fall back asleep. 06h00 reveille is sounded. We load up and wait for the sun to rise for better visibility and head back to South Station. 08h00, we reach base; unload, and I head straight to my cot to catch some more shuteye. 08h45 I have awoken again, feeling like that ole train had hit me again—this time, I had to get up to attend a ceremony for Gleasons promotion to Lance Corporal. After congratulating him, Cruikshank approaches me, stating he cannot hear anymore from his left ear. I grab my medical pouch and flush out his ear with clean water and eardrops. I got on the sat phone and called home because it was now too hot to go back to sleep after I got off the phone. 1st SGT advises me we are going on another mission. It seems the work was endless here these past few days. I was pleased that it was making the time go by swiftly. The only thing that sucked was the sleep deprivation. 10h30, we load up on the pigs, this time, I am riding with SGT Abelar and the company commander on the Black-6 vehicle, which was a LAV25 variant with a turret. I haven't ridden on one of these since my tour in Iraq, so it was exciting for me until the dust started blanketing us the whole way back to whites' position. They called up another IED 9 line. We reached them at 11h30, got a brief with EOD, and we started our dismounted patrol towards one of the villages to the

IED sight. It was noon and hot outside. The patrol was interesting; I got to get a close-up experience with another EOD operation. We had to cross through a kilometer of swampland. Squishing through, my feet and boots were soaked and covered in mud. The land smells in resemblance to that of a muddy football field and donkey/sheep/camel shit. After about 45 minutes of walking through the swamps, grass, bushes, and dried mud lands, we reached our destination just outside a farm village fenced with palm trees and cornfields. There were 2 intersecting dirt roads that we set a security cordon on, and EOD techs went to work. 14h00 EOD finds the core piece of an old battery and nothing else, so we head back to the same way we came crunching and squishing away back on the path to white HQ B Section coil. 14h45 we reach back to the coil. My shoulders are feeling nice and burnt out, and I am covered from head to toe with muddy sweat, and my feet are soaked. After being de-briefed, we load up on the pigs and head back to South Station, getting powdered with dust along the ride back. 15h30, we reach South Station and head to the tents. I strip off all my clothes and give myself a baby wipe bath. 16h00, feeling exhausted and ignoring the heat of the day, I pass out in my rack. 18h00, I wake up because my body completely sweats out all its hydration and because I felt a mouse run across my stomach. I rush to a bottle of Gatorade and slam it down my throat. I get dressed in my dried-mud-crunchy-cammies and head to the lance tent to see Cruikshank flush out his clogged ear out again. It was now 20h30, and I was contemplating if I should go to the gym or catch up on sleep. 21h00 I decide to go to the gym and then pass out feeling satisfied with my workout at 22h00.

October 7 – 06h30 reveille. We head back to the pigs and load up on Black-6 again; we waited for 2 hours because the Afghan police were moving slowly as hell to get ready. 09h00 White Platoon reports contact over the radio. They had another rocket fired upon them. The CO says, "Let's roll out!"

And we race out, dropping off gear at Red's VCP station, then rush over to White's coil. 11h00 as soon as we arrive, we hear and see bombs falling from the sky followed by smoke clouds. We called for an airstrike of cluster bombs to rain down upon the position from where the rocket fire came from. We called in six strings of fire missions called 'shake and bake,' which was a mixture of High Explosive and White Phosphorus. It was a spectacular sight right behind the hill in front of me. Noon, the fire missions ended, and the CO needed me to be the vehicle commander of the MRAP EOD vehicle. I enjoyed the air-conditioned ride back to South Station. 13h00, we arrive back at South Station, and I head to my tent to relax. I fell asleep ignoring the heat and dreamt about flipping over in a LAV and pulling out injured Marines. I probably had these dreams due to thinking about what I would do in a situation like that every day while rolling through. I awoke at 1800 feeling very dehydrated and sat around watching movies and relaxing in the tent.

October 8 – I wake up at 09h00, and King shows me the mouse he caught and is keeping in an empty water bottle as a pet. The shit we did for entertainment during these times never seemed to find a limit on extremes. I go to the gym at 11h00, then SGT Hernandez shows me his pet mouse, and I sit around all day bored and sweaty. I slept pretty much the entire day today, which I am exceptionally happy about. The only interruption was being woken up for having to give a band-aid to a Marine for a blister on his foot. A lot of Marines would question my sleep patterns and abilities, but my reasoning was that the more I slept, the faster the boring times on deployment would end. It turns out it was working. As opposed to Iraq, sleep really helped the time go by quicker. Plus, I was always being woken up for something, so sleeping at any time could be very valuable. I woke up at 20h00 and started wandering around bored again. I would wander around the base looking for different wildlife creatures for entertainment. Watching the same movies for the third time was getting redundant.

October 9 – I wake up at 10h30 and hear the sounds of Charlie Company vehicles pulling in on the ramp, and I feel relief because that hopefully means were going back to Payne tomorrow. I went to the gym at noon then sat around until 16h00, yet again fighting boredom. I then took an inventory of the medical bags on the vehicles here then sat around for another couple of hours. I went back to the gym, and SGT Abelar rushed in a while in the middle of my workout and said, "load up; we got to go." Finally, some potential excitement. So, I rush to the pigs, gear up, and load up. No one seems to know what the hell is going on. I am nervous sitting low in the vehicle, preparing my mind for the worst situation possible. While en route to wherever we were headed, I asked SGT Abelar what we were doing. He replied the weather blimp spotted several men digging in the ground west of our base and we were teaming up with Jackal (the British Royal Marines) to intercept. Those guys were pretty badass. They had their accents and their rules of engagement and different weaponry and would say some pretty obscure things that we would always find hilarious. It's Black-8, Black-4, and Black-6 out here, and all vehicles get stuck, and we spend hours trying to dig ourselves out of the wetlands. By the time we broke the vehicles free from the mud and soft wet sand, the mission had been canceled due to losing sight of the IED diggers. Then we returned back to South Station. Well, that was shitty. 00h15, we reach base, unload, and I head to bed.

October 10 – 06h30 reveille. I go out to the ramp to help the Marines fix the vehicle that was malfunctioning during trying to recover each other last night. I went back to sleep after an hour of that and woke up at 11h30 and sat around listening to music on King's iPod for several hours. 14h00 rolled by, and I decided to drag my ass to the gym again. I start my workout as usual and go to the corner of the room to use the pulley, and there I see a small ball of quills nestled up in the corner behind the weight. I take a closer look, and I can't believe my eyes. It is a hedgehog! Just like the ones I had when I was a kid. I was so happy to find one out here. This gym wasn't a safe place for him, and of course, I am a guy who loves animals, so I pick him up, and he is a friendly one and the cutest little thing in the world. I was overwhelmed with joy. I named him Hedgie, just like the first one I had. I quickly carried him to my tent to show SGT Abelar, and with excitement. We took pictures and videos holding him, then I showed him around to the Marines and talked about his species like it was a show and tell day at school. I brought him back to the tent and collected some bugs, and he ate right from my hand. I laid in my rack for an hour and watched him crawl around in my bug net around me. I noticed he had a tick on his ear, so I used my plyers from my Gerber to remove it, and it was a cinch. Now I love it here because I have a new friend to keep me busy. I laid around watching Hedgie crawl and snoop around. Then, at 19h00, I decided to go back to the gym then looked for bugs to feed Hedgie. I found a mole cricket which seemed to be his favorite. Orta tells me I have a package from a girl back home, so I went to grab it from the pig and was very pleased and thankful. 23h00, I put Hedgie in his box and crawled into bed.

October 11 – 08h00 reveille. The Marines and I wake up cursing the fact that we have another EOD mission again. I noticed that Hedgie escaped and was nowhere to be found, but I had more important things to worry about. I was glad for his visit and happier that he was free to roam around again instead of being trapped in a box like I was. We load up on the pigs, and I get on Black-8 and head to Reds' position. We arrive at their outer coil at 11h00 and load up with 3 EOD members and patrol deep into the village to link up with the rest of the platoon holding position at the IED site, near the newly built VCP here. Surrounding us are farmlands, cattle, sheep, goats, dogs, children, crops, and trees. The CO debriefs us before the dismount and directs me to remain on Black-8 and establish a CCP (Casualty Collection Point). I set up my tarp for shade and stand by keeping an ear on the radios. Members of the Afghan Army are here to assist us with our operations and establishing security at the VCPs. I see Marines from Red Platoon and ask how everything is out here, "Shitty." they all grumble the same one-word answer, they ask me the same about South Station, and I reply, "Shitty." Everyone's morale has plummeted as a result of being out here for so long with little to no rest and constantly operating. It makes me worry about how we are going to perform on our upcoming Bahram Cha operation if we don't get some rest or a morale boost or something. Not everyone's head is going to be focused, and that is a dangerous thing. After an hour or so of sitting around, the patrol returns at 13h00 and loads up. We started our second mission, which was to investigate a possible suspected weapons cache. We drive for 2 hours through mountainous terrain. In several areas en route to our destination, the views of the river valley were spectacular. We reach a very small village comprised of some mud huts, cattle, women, and children. The males were on their tractors in the crop fields. We rolled through the tiny village and parked right next to the crop fields. The dogs were barking, and the people were looking at us as if they had just seen a ghost. The scouts and EOD punch out to investigate for any signs of a weapons cache, and this goes on for about 3.5 hours while I stay in the cas-evac vehicle standing by in case of any emergencies. It is now 17h30, and the sun is setting behind the mountain tops. Another hour goes by, and the CO calls off the search. 18h30, we prep for night load up and head back out to South Station. We arrive at 22h00, and I go to sleep, ignoring the fact that I was still covered in a cake layer of dust.

October 12 – I had a pretty cool dream about being a splinter cell spy and awoke at 11h00. I received a package from a friend from back home with food and some mouthwash. I wrote her a thank you letter and sat around sweating my balls off for the rest of the day.

October 13 – I woke up, and I was alone. The rest of the platoon had gone to Payne for the 25mm cannon range. I sat around all day sweating like crazy, doing nothing. Today is the birth of the USN. The Marines returned at 21h00, giving me a hot tray of cold shrimp, lobster, and steak, and said, "Happy birthday, Doc." 23h30 after lying in bed, the corporal on guard came in and notified me that there was radio traffic for me in the COC. I get to the COC with a sense of emergency and roger up. It is Doc Hermon on the other end of the transmission. He reports to me he has a Marine away from his current position experiencing shivering, feeling hot, nausea, body aches, headaches, dizziness, and trouble breathing. I tell him to get me a set of vitals. He takes a team to patrol to the sick Marines' position and rogers up. 30 minutes later with vitals of 104.7 temperature, 105.4 core temp, pulse 120, tachycardia, unknown blood pressure due to a broken cuff. I roger up and tell him to stand by for a cas-evac. I call over to BAS at Payne to give heads up on receiving a patient. The MO tells me to hold off on the cas-evac, administer Tylenol, wait 1 hour for signs and symptoms to increase or decrease then administer Ibuprofen if Tylenol has no effect. I also have to get the patient history and any other information and report back. An hour went by, the Tylenol had no effect, vital signs remained the same, but now the patient is starting to convulse and having more trouble breathing. I make the call to cas-evac the Marine. The Marines in the COC send up the proper 9-line report under my guidance for a respirator in the special equipment. The CO ordered HQ quick Reaction Force Bravo and myself to convoy to Red Platoon's landing zone (LZ) in case the chopper cannot land.

October 14 – 02h30 We load up and rush out to the LZ! Halfway there, the chopper successfully completed the mission, so we headed back to South Station and got some rest after arriving and me attempting to call and debrief chief at 03h00. 09h00 I wake up and go to the gym, then hang out all day until 16h00 and give a CLS (Combat Life Saver) class to my Marines. I fell asleep at 19h00 due to another wave of boredom.

October 15 – Nothing to do all day long; it is just hot and miserable. 17h00 I stepped into the COC (command operating center) to see what was going on, and I overheard radio traffic of an Afghan Border Police (ABP) patrol hitting an IED at Reds position, killing one member of ABP; he lost both legs and an arm.

October 16 – The PX truck came to South Station along with a resupply of chow and water at 08h00. The PX truck was a 7-ton vehicle designated to transport typical gas station-type amenities such as tobacco products, energy drinks, magazines, candy, and other morale-boosting goodies. I thought it was going to be another day of sitting around doing nothing, and that is how it was until 11h00 when I was called to load up on Black-8. Noon we roll out up to Payne only to pick up 2 Marines that were handling the detainee that was captured at Red's position after the IED killed the Afghan Army member last night. We traveled to Red Platoon and arrived at 15h00 to the inner coil. I got the bum scoop on the graphic event that took place there, and every detail was so gruesome. According to CPL Scotty VanTheemsche, "A unit from the Afghan Army was patrolling outside of a village near their position. The sun was starting to go down. Suddenly, BOOM! Everyone was like, what the fuck!? Apparently, this Afghan Army member stepped on top of an IED, and body parts and blood scattered everywhere. Another member of this man's team was holding what was left of his foot still in the boot laying at the blast site, crying and rocking back and forth. We rushed to the scene as fast as we could, and others provided security. Immediately after the blast, other members of the Afghan unit panicked and were firing their weapons out in all different directions toward the village. We took cover, and LCPL Navarro and another SGT started working on this blown-up member. They managed to slap three tourniquets on the limbs that were missing and began to transport him out of the area. He ended up staying alive for thirty more minutes. As the rest of the platoon and I approached, I could see body parts scattered all over the field. It was a severely fucked up sight to see. What was even more insane was that once EOD arrived, they uncovered a secondary pressure plate IED at the T in the route. We could have all died. The only thing that saved us was the electronic signal blocker we were carrying. It blocked the signal of the detonator while we were trying to save this Afghan Policeman." As we helped the platoon recover and further investigate and report the event, I pulled out the chew rope, and doggy treats that a friend from back home sent to me to give to Molly. She loved them. Molly and I played tug of war with the rope, and she inhaled the beef stick treat. 16h00, the CO gave the order to prep for a dismounted patrol into the village to talk with the elder about the event last night. We gear up and head into the village, stopping at one of the mosques after crossing several crop fields. The mosques were the sacred temples where the villagers would conduct prayer which was also where the loudspeakers were located that played the eerie prayer we heard every evening. The Marines and I post a security perimeter around the area while the CO talks with the village elder. I kneel against a short wall made of mud surrounding the mosque and scan the area. After about an hour of scanning, some village children cautiously approach me. There were 5 girls and 3 boys aging from 2 years to 6 years old. Once I give them a smile and a wave, they sense no danger and begin saying hello and touching my gear that I was wearing, speaking their native language asking for things like water, money, and candy, mainly chocolate; they always ask for chocolate. One of the young boys even says the English alphabet. A little girl touches my arm and feels my morphine pens in my pocket, and tries to reach her hand in to take them, but I grab her hand and say NO! She smiles and starts mocking me and repeating everything that I say, and laughing. More kids start crowing around me. They were covered in dirt and torn clothes. I felt as if I were about to get jumped by a gang of children. In order to keep safe from getting my gear stolen or becoming complacent, I told them to go away because I needed to stay alert, and they were obstructing my vision and sectors of fire. I wish I had some candy to give them but whatever. I told them to shoo and made the hand gesture to go away. They left laughing and smiling, and I felt happy for once during this operation. I was happy to be here for the children and to provide them with efforts for a potentially better future. Another ten minutes rolls by, and we head back to the pigs at the VCP, passing herds of baby sheep and baby goats along the way. 17h30, we load up and head to Camp Payne and get some good chow, and I head up to the BAS to check in with Chief and HM1 and give them the rundown on how the mission was going. Shortly after, I headed back to the ramp where all of the vehicles were staged. I met members from Weapons platoon, and we were very happy to see each other. I helped Orta grab the mail and load it up on Black-8. 20h00, we headed back to South Station. Today turned out to be a good day because we did something productive. It was difficult, however, to try not to think about the gruesome event that took place the night before. [Just another thing to add fuel for disturbing future dreams.] There was no time to let the mental pain and anguish set in because there was a mission that needed to be achieved here. Suppressing negative thoughts and mental images was a trait we had to learn in order to press on and continue the fight against terrorism.

October 17 – I woke up at 08h00 and went to Black-8 to change out my broken stretcher with a new one. The unforgiving rays of the sun here would tarnish the materials and would have to be replaced often. After that, I went to the gym for an hour. Feeling lonely again, I wanted to talk to someone back home on the satellite phone, but since someone violated operational security about our upcoming mission, the battalion put everyone in river city status, so the only line of communication to back home was from letters. At least until our last mission. I spent the rest of the day watching the last episodes of scrubs. I can remember just lying there for hours upon hours, only getting up to use the bathroom, sweating my balls off in the tent, trying to stay sane, and trying to keep my mind occupied. 22h30 rolls around, I decided to go to the gym yet again; I spent an hour there thinking that I was going to come home looking like Vin Diesel, but what I actually was doing was burning so many calories in the Afghan heat that it was making me look like Jack Skeleton. I watched some more episodes of scrubs until I fell asleep. That was the jist of my day. I spent the majority of my days at South Station- eating, masturbating, watching reruns over and over, masturbating, working out, masturbating, going to the bathroom, masturbating, and then finally falling asleep in the perfuse heat. 05h30 reveille. Word comes down the line, and we load up the pigs. Black- 8 is filled with chow and water to resupply Charlie Company. We head to Payne to load up with supplies, then back to South Station. This took us to roughly around 10h00 at this time we headed to Red Platoons position and arrive at 10h30, where we unloaded the chow and water and headed back to South Station where I stayed in my tent and tried to keep my mind sane until 17h30 when we pushed back to Payne for unknown reasons. Nevertheless, we arrived at 17h45, where I had an hour and a half to do whatever I wanted. So as a good Corpsman, I decided to head up to the battalion aid station to grab HM2 Delacruz and go to chow. Shortly after, we went back to the BAS, where I saw Doc Donny and conversed for a few minutes until it was time to leave back to South Station. We head back and arrive, unload, batten down the hatches, and lock up the pigs. At this point, I continued the normal routine with a full stomach, watching scrubs, thinking of loved ones back home, and contemplating whether I wanted to go masturbate or go to the gym. I decided to rest tonight instead and that was the extent of my day. One exciting thing happens, and the rest of the day, I must keep occupied any other way I can. Some would say that it was like being trapped in a small sand box with walls built of sand called Hesco barriers and serving a prison sentence within. The next day, I woke at 08h00 to the sound of 1st sergeants' voice telling us to clean the tents and shave because the XO's of 1st and 3rd LAR battalions were paying a visit. As the Marines clean, Orta finds a baby kangaroo mouse under his rack, and he screams out excitedly, "HOLY SHIT! IT'S A TINY KANGAROO!" But much to his dismay, it was only a certain species of a mouse with amazing hind legs and jumping abilities. It was still funny to watch him freak out about seeing something like that. At around 08h30, the XO's from the different units showed up, and everyone went about their business while they take a tour around the makeshift forward base. They must have been thinking to themselves along the lines of "Shit, this place sucks ass." I took a four-hour nap, then we got the word to load up to go to Payne for unknown reasons yet again, and then we came back. 20th October was our last day living at South Station; I was so happy to find out that we wouldn't be living here anymore; boredom was finally over, or at least wishful thinking. Even though word was passed that we weren't going to be living here anymore, I didn't fully believe it, I mean, this is the USMC, and word ALWAYS changes. So, my excitement was suppressed to a mere disbelief until it actually happened. With that being mentioned, today was going to be a day of packing. All the Marines packed up all their stuff with excitement and relief and loaded up the pigs with all of our belongings. We packed until it looked like we were never there and then slept off the long day. This was the end of the nightmare at South Station.

October 21 – Reveille was at 05h30, first thing was to hygiene, stuff our faces with shitty MRE food, and then load up on the pigs and start pressing towards Camp Payne. We cleaned the vehicles at the wash stations and did regular maintenance on the ramp. As the Marines did their duties working on the vehicles, I did mine and headed to the BAS to meet with the Chief. He gives us a brief which is more or less just a pep talk about how proud he was of us and how we will be making history, and that we haven't done an operation like this one coming up in seven years. At this point, I felt proud of myself, but Chiefs' word of significance about being so important kind of worried me a little bit. All the old stories of the huge battles in history were running through my head which made me think I am probably going to die, but it was an awesome feeling that I will probably go out in great glory. This is something I have always envisioned myself doing; being the bad-ass and saving Marines in a glorious battle. After Chiefs' pep talk, my company had to go sight in their rifles at the range. I hated doing this because it was a pain in the ass standing out in the hot sun all tacked up with hot gear on, shooting targets, hearing other people run their mouths about others bad aim, but at the same time, I understood the importance of being on target for battle. After shooting at the range, we head to guess where chow! I finally got the chance to take a shower with running water. This shower made me feel like a billionaire. These showers were a replica of a trailer with shower heads sticking out the walls. Even though the shower floors were soaked with mud and dust from all the Marines, I remember stepping out of the shower and being covered in dust immediately, but at least the layer of gunk wasn't as thick because I had just cleaned off the old 20 layers from my skin. The next several days are going to be preparing for our final mission. After chow, I lay my head to rest on an actual box spring mattress with the cool air of a dusty old air conditioner blowing on my face and felt like I had just won the lottery.

October 22 – This day was a day of progress and preparation as anticipated. We got stocked up on medical supplies, and the Marines did combat maneuver readiness and got issued enough ammo to supply the entire United States twice. I received a letter from my best friend James and one from a girl back in California, but most excitedly, we got our brief about the mission. Finally, all this hype up was disclosed to our knowledge. It's going to be a rain of absolute hell for the enemy, and I can't wait. In short, 23 hours of complete destruction and annihilation of a certain location. Later that night after the brief, I went to the gym, and then I went back to the tent and found a bunch of the Marines talking about Wagoner and about how he was sleeping and how he falls asleep so easily. For instance, he once fell asleep on the top the vehicle while it was moving, getting pelted in the face with rocks and dust, with his mouth wide open waiting to catch flies for extra protein. One time he was sleeping on his rack, knocked out, so the Marines decided that he must get the duct tape. The Marines were making bets whether he was going to wake up or not while getting duct-taped to the bed. Low and behold, mummified to the bed, he was still in a deep slumber. After everyone stopped paying attention and lost interest, he finally awakens, and all you hear is, "AWW, YOU GUYS ARE ASSHOLES!!" as I laugh myself to sleep listening to him struggle to escape. The next day was another day of production as the Marines did the entire vehicle maintenance and preparation. While the Marines took breaks, Doc Donny and I gave medical classes to them. We retained our normal chow routine, and that was what the day consisted of.

October 23 – Woke up at 06h30 and went to chow. After that, I headed to the ramp at 08h00, where the Marines were there preparing their vehicles and their personal protective equipment for inspection at 13h00. I didn't really feel like doing maintenance on the vehicles because it wasn't my job, so I took this opportunity to visit the BAS for last-minute medical supplies. This is where I "Skated" until inspection time. They always called me "Skater" because I was always getting out of doing maintenance but was still doing my job. I would talk shit back to the Marines, telling them to come up to the BAS and conduct sick call with me. We would talk more shit back and forth to each other until I ended it by saying, "Okay, I'll remember this conversation as I'm working on stopping you from bleeding out in the field." They would then reply, "That's fucked up, Doc; I'm just kidding." They were good times, and we all ragged on each other constantly like brothers. When it came down to it, we would die for each other in an instant without a second thought. I was obligated to participate in this inspection, in which case the day was pretty much over because there wasn't anything further to do. I went to my hooch and took a nap until evening chow, then went to the gym while everyone chilled out at the hooch. Tomorrow is the day that we leave for our last operation at Bahram Cha. I have feelings of excitement, but I am also scared for my life; this place is filled with terrorist training camps, drug and bomb creation facilities, Taliban headquarters, Al-Qaida members, and many others who wanted us dead. As I lay in my rack thinking at night of what the future may hold, I try to coax with my thoughts of fear and mentally prepare myself for a massive casualty experience while I slip off into nightmare land.

October 24 – 07h00 reveille. I woke up and headed to chow, wondering if this was the last DFAC chow that I would ever eat. After chow, we all gathered at 09h00 at the ramp to get communication checks, maintenance checks, and last-minute PPE inspections. I ensure all Marines IFAKs (Individual First Aid Kits) and tourniquets are operational, and at this point, we sat around on the vehicles for a short while until the entire task force involved in this operation gathers at the center of the ramp for a brief from the battalion CO. His words were moving and enticing as he rallied the troops and ensured us that we were about to set forth and make history.

We all shouted in glorious union and reckless abandon after the speech was over. The time to kill was approaching. We could all feel it deep within our spirits that this was going to be epic. 11h30 was when the brief ended, and everyone rushed to chow, yet again that this was the last DFAC meal I would ever get to savor. We got back on the vehicles after chow, and yet again, we get com checks with each vehicle and each radio and hung out on the vehicles some more. One of my Corpsmen comes to my pig with some field and stream magazines; we sit and discuss our upcoming mission and go over casualty protocols. At 12h45, we load up and stage every vehicle for movement. At 13h00, Doc Herron comes to my pig, where we also discuss casualty protocols while waiting for the next order to be given. Mike Hernandez then brings Molly over. She jumps in the back of the pig, and I play with her for a while, awaiting the next brief or prayer from the Chaplain. At this point, I was becoming more and more nervous. At 14h00, the chaplain comes over, and Bravo Company forms a school company around him as he leads us all into prayer. We all bow our heads and pray to our Gods to watch over us as we commence battle with the enemy. Shortly after, the company CO gives his speech. I then go over protocols with my Corpsmen and give my speech. I can remember the CO's words, "We are making history this week gents, today starts the day where the world will forever know of a great battle against the enemy, and we are part of that." Master Sergeant comes to the center of our crowd as it sits in silence for a moment. The silence amongst the crowd was broken as he shouted, "Go red count one!" Which means starting the vehicles and prep for movement. We all disperse with the famous battle cry "Hoorah!" and head to the vehicles, mount up, tack up, and clench our fists and teeth and lower our brows with a beastly look of focused determination. We were all war savages. We were all ready to die for our nation in order to eliminate the enemy. Communication is a go, vehicles are a go, and the

Marines are blood-thirsty. You can tell how blood-thirsty they are by their actions and motivation, screaming Hoorah at each other and acting like they were entering the field of the Superbowl. We were all ready to do something meaningful after such a long time. It's time to go kick some fucking Taliban ass. All the vehicles depart friendly lines in proper order of movement. The long convoy south has finally begun. Our convoy drove for a total of four hours, only stopping for hot checks for several minutes along the way. As we continued our journey, we passed the usual travelers with water and fuel jugs and tractors with mountainous piles of sticks on the trailers. At 18h00, the entire company stops and sets up a coil due west from our target. This was to ensure sort of a decoy operation that we were conducting, which was consisting of a normal convoy patrol. I hop out of the vehicle and instruct junior Marines to dig saddle trenches and piss pits along with a burn pit for trash outside of the coil. I hop back into the vehicle and check my EDL consisting of serialized gear, weapons, night vision, and bayonet. As I am checking my EDL, I can hear 1st sergeant give a Marine an ass chewing for not wearing his PPE. At that point, it became clear to me that we were always to remain tacked up. 19h00 rolls around, and I crack open an MRE. I ate a little bit and took my doxycycline before laying my head to rest. I slept for several hours but woke up every 10 minutes feeling cold and nauseous. My body wasn't used to how long it takes MRE's to break down in my system, and the doxycycline, as usual, made me feel like shit, and this went on as a continuous pattern until 02h00.

October 26 – 02h00 Doc Dony wakes me up for radio watch. As soon as I stepped out of the pig, I vomited several times, waking up several Marines. They came over to see if I was alright. After puking for about 5 or 6 times, I stood an hour of radio watch. It was so cold; the temperature had drastically dropped within the past week. After an hour of sitting in the C2 radio vehicle on watch, I crawled back into bed, waking up every 10 minutes feeling nauseous and freezing. 08h00, the Marines awaken and start their normal morning routine of hygiene, coffee, and yelling at each other for not wearing all of their PPE and also making fun of the fact that I was puking my brains out. I feel as if I have a hangover and am still nauseous, so I take a Phenergan to calm my stomach. It is 08h30, still cold as balls, and I'm sitting down in the pig waiting for the medications to kick in. I lay my head down for several hours and awoke at 13h00, and immediately threw up again. I still feel like shit. Doc Dony gave me some medications to help clear out my system because we think my digestive system is backed up. The Marines run stretcher drills as Dony, and I provide guidance and constructive criticism. I keep pounding water, waiting for the medications to kick in. All of the Marines around me are practicing for combats' worst-case scenarios. The Marines also made a terrain model using rocks and dirt to represent our mission. MSGT goes over the mission step by step and explains everything, 'breaking it down Barney the dinosaur style' or so as the Marines would say, so even the dumbest Marines are on the same page. I feel our biggest threat is going to be enemy personnel shooting weapons systems at us from the mountains behind us and IEDs in the ground. I fell back to sleep and was awoken to stand 02h00 to 03h00 radio watch at the C2 again.

October 27 – 02h00 I immediately had diarrhea after being woken up for watch but felt a little better due to getting rest. I was relieved by Doc Dony at 03h00 and went back to sleep. I slept in until 09h45 and finally ate something. I still feel under the weather, but every time I slept, I felt better and better. 10h30 Doc Rey comes over to grab some medications for one of his Marines with diarrhea. I lay back down to take another nap. 12h30, one of the Marines comes over to get a band-aid for a cut on his finger. As I was cleaning it, he suddenly convulsed, passed out, and started twitching. I pulled him into the pig and propped him, sitting up using my body for him to lean on. I called over to Dony for assistance. He brought over a nifty glucometer to test his blood sugar. It was low along with his core temperature, so we took his vitals and instructed him to pound water and stay in the shade for the rest of our time here after he came to. At 12h45, we pack up camp. 13h00, we start our long movement south to the Logistic Supply Area (LSA). In order of movement, all vehicles stop at a small hill to test all weapons systems. We drove south through the desert, passing numerous sand dunes and open flats. We stopped and coiled up at 17h30, about 35 clicks (Km) north of the LSA. We would have continued, but personnel was not ready to receive us yet. The sun was setting past the mountains ahead of us, and the temperature was beginning to drop. I hop out of the Black-8 to check on the Marine who had low blood sugar earlier. He seemed to be doing fine. My body aches have gone, but I was still experiencing stomach discomfort. We held our position at the coil for the night. I was awoken to stand watch at 23h00 to midnight. I stayed awake until 03h00 experiencing explosive diarrhea. It was a shitty way to start this big mission, to say the least.

October 28 – 06h00 reveille. We pack up, tac-up, and wait for the next order. I am feeling much better, other than having bubble guts and being tired. 07h30, we push out to the LSA, passing numerous large mountains along the way. At 08h30 we reach the LSA, refuel, and stock up on chow and water. Just as we stage our vehicles inside the massive coil of Marines, a thick sandstorm hits. Sprinkles of rain and heavy winds rumbled on top of us as the air turned red. Visibility became very low. The sky was a thick orange that carried dust and rain. We parked next to Tien's vehicle, and I was happy to see him. We conversed for a while as the violent wind continued to shake the vehicle. 13h00, we move back to our previous coil that is now in a closer location to the LSA interdicting vehicles. We were rotating sections of the company to be topped off with fuel and supplies. Thunder clouds overhead make familiar sounds in the sky as the wind continues to blow with rage. It appears our raid has been pushed back 24 hours. I took a nap and awoke at 17h00. It rained hard for a good ten minutes on us. After Dony and I passed out the overload of pole-less liters to whoever didn't have one, we headed over to Black-8. I cracked open an MRE and made fun conversations with 1st Sgt and Orta, and just then, a thunder and lightning rainstorm smashed down on us. As I finished eating my MRE, the rain stopped, but the thunder and lightning kept going. As we listened to the thunder rumble and watched the lightning through the periscopes, we talked about random things from back home. We were sitting there in the vehicles with all hatches closed in what seemed to be the middle of nowhere, listing to the vehicle get pelted by the storm on the outside, feeling it shaking and rocking with the gusts of wind. If the weather continues the way it has been going, this mission might get canceled. The rain started again as I lay my head to rest.

October 29 – It thunder stormed throughout the night. I awoke at 06h30, and I stepped outside the vehicle to take a piss. All the sand was moist, and everything on the vehicle was clean from the rain. I geared up and walked around the coil to check on my Marines and see how they made it through the night. Everyone was good, but most complained of bad sleep. They had to find uncomfortable spots in their vehicles to cram into for shelter from the storm. No one came prepared for thunderstorms on this operation. I came back to Black-8, did my hygiene, and looked for something to eat. We did nothing but kick sand until 09h15. At that time, we loaded up on the vehicle and escorted the company CO back to the LSA for a meeting. The weather was fair today. If it keeps fair, this mission should happen tonight. 14h00 air shows on station. MV-22B Ospreys and Blackhawks arrive at the LSA. 15h00, the entire company links back up in another massive coil outside the LSA. 16h00, we get the order to begin our movement towards our first objective. The raid is finally beginning. Everyone is excited to do what they have trained for their entire military careers. Bravo Company stages in order of movement, and we push further south towards Bahram Cha. 17h00, we link up with the rest of the supporting elements for this operation. We stage and shut down the vehicles. To the right, the sun was setting behind the mountains (crocodile ridge). It was beautiful! To my front, as far as the eye can see are parked vehicles, MATV's, LAV's, MRAPs, 7-ton etc. all were ready to wage forth into battle. Doc Herron came over to get some medications from me, and he said most members of his platoon have the shits. What a bad time to get the shits. We continue to await the order to push out. The feelings I am experiencing were similar to those of when we raided Derzay. 18h40, the Marines all load up and head out. I don my NVGs and cautiously scan left and right. We drove into a lightning storm. Overhead I could hear jets and choppers traveling south to drop bombs on our objectives about halfway to our staging point. It rained freezing droplets for about an hour during our movement. Those who were popped halfway out of the vehicle were soaking wet and freezing. 21h00, the rain stopped, and 21h45 explosions started erupting in front of us, and just like lightning, bombs were lighting up the skies behind the mountains. We reached the mountains outside of our target village around midnight.

October 30 – We drove through very dangerous channelizing terrain. We were driving between large hills towering over us, listening to the sounds of bombs dropping nearby. Our eyes were peeled for enemy ambush upon the ridgetop. Lead vehicles were firing MICLICS (rockets with strings of composition B explosive) to clear the route of IEDs. Exciting but nerve-racking. 01h30, we reach the other side of crocodile ridge, where I witness demolished and still burning buildings. C130's fly overhead with deep, deafening rumbles and unleash the Gatling cannon several times from the sky. The sound of US firepower combined arms was spectacular and would make anyone feel proud to be an American. I was watching a barrage of metal and flame rain down from the sky at the enemy target that would leave nothing but a pile of rubble and smoke. 05h30, we finally reached our objective. The sun was rising yet we were still cold as hell and tired. We observed and continued to carry out the mission. Explosions and gunshots erupt from our cannons as we continue to demolish the city. We set up an area for the CCP and MCP. Right in front of us was the sight of burning buildings, smoke rising in the sky, and the Pakistan border. Surrounding both us and the city stood large mountains. Bahram Cha is like that of Derzay. The only difference here was more vegetation and many more buildings very close together. 07h00, we shoot tube-launched, optically tracked, wire-guided (TOW) missiles in the city wall, demolishing it to rubble, and Blue platoon floods in to infiltrate like a pack of mad dogs let off the leash. There was a big sign made of white rocks in Arabic writing that we could all see from our positions. It resembled that of a college letter printed above a town. 30mm Gatling guns fly by several times and destroy it. It was a display of my favorite weapon of all time, the A-10 gunship.

Witnessing the A-10 gunship was a devastating display of force through firepower. Through the years, this aircraft had been equipped with an array of 30 mm GAU-8/A cannon; up to 16,000 pounds of mixed ordnance on eight under-wing and three under-fuselage pylon stations, including 500-pound Mk-82 and 2,000 pound Mk-84 series low/high drag bombs, incendiary cluster bombs, combined effects munitions, mine dispensing munitions, AGM-65 Maverick and AIM-9 Sidewinder! The A-10 was known as "Warthog," and it was fitting. Imagine an aircraft with so much precision and firepower that when unleashed, the sounds could only be described as ten thousand dirt bikers revving their engines to the max while each wielded chainsaws and threw several grenades all at once for good measure. The sequence of its attack when witnessed was that of a large aircraft breaking through from behind the clouds in the sky heading straight toward its target. The rounds were unleashed as they continued to approach. Then a pause. Then the sight of the rounds hitting the target, making very similar sounds as the rounds explode, disintegrating its target to pebbles and pieces.

The enemy returns fire with a single rocket shot that exploded 300 meters in front of our faces. We returned with a barrage of ground fire consisting of a 25mm chain gun, M240 coax mounted machine gun, and a TOW missile as a "Fuck you" message of screaming bald-eagle hate right back at them, annihilating the point and everything in the vicinity the enemy rocket shot came from. The fighting continued like this for several hours. It began to die down as the enemies were exterminated with every ammo dump from our end. 08h45, the only sounds heard now is radio traffic and overhead aircraft. I sat low in the vehicle and closed all the hatches, and listened to radio traffic of the Marines operations going on around us, praying to God that everyone else was alive and well. 09h05; Another explosion to our west side erupts. Then it dies down after we yet again return fire. We go into snowstorm status and continue to observe activities as they unfold around us carried out by all the different combined unit activities. We sat again in silence, waiting to progress to objective 2. Explosions and gunfire are heard from the other side of the hill that Alpha Company was occupying. Blue platoon reports 50 plus women and children egressing out of our Operational area towards Pakistan. If we had seen anyone outside their homes today looking at us the wrong way, we fucked their shit up (Excluding women and children). It's fucking game over for the enemy. 10h15 now 150 plus women and children were reported by Red Platoon of trying to cross into Pakistan away from this city but were denied at the border patrol checkpoint and said fuck it and started hopping over the berm off to a distance from the checkpoint anyway. 1st Sgt directed Orta and I to catch up on some sleep, so that is what I did for about 3 hours. [You are probably thinking to yourself right now, "How in the fuck can anyone sleep in the middle of something like this!?"] I awoke to the piercing sounds of explosions and helicopters swarming overhead at 14h40. The battle continued to rage. 15h30 explosions continue erupting in random locations all around us, but for the most part, things were slowing down here at the CCP. The taskforce was still currently clearing out the bazar, and we were awaiting our next orders like thirsty savages with the urge to keep fighting. 16h00, we escorted the Company CO Hylander-6 vehicle closer to the bazaar while providing fire support and security. From here, I had a clear view of the entire city. I broke out the binoculars and scanned to see miles of destruction and burning rubble. We came here to fuck their shit up, and we did just that. 17h00, we return to the CCP. 17h30, the CO did net call to talk to all platoon commanders, "Mission complete, men." He gave the order of egression back. We continue to sit and wait for the word to push out. Night falls, and I take first watch to let Orta get some sleep before the long and late drive back to base. All was silent until 20h00. We spot several men in the mountain in front of us carrying rifles and RPGs. Other teams confirm the sighting, and we open fire, blasting the 25mm chain guns at them. It was death on impact. It scared the shit out of me because I wasn't listening to the radios at the time, and all of a sudden, I heard gunshots. Watching the rounds fly through the air and explode at the enemy's position was quite a light show. Body parts became airborne. 20h40, we spot personnel digging IEDs. Controlled detonations erupt from within what is left of the bazaar. I decided to get some rest after Orta woke up and relieved me from watch. As I slowly drifted off to sleep, I could hear over the radio Marines spotting personnel carrying more RPGs and other weapons and the sounds of our 25mm cannons blasting them away. There were still bad guys in the area, but they were late for the party. Little did they know we were still willing to give them a slice of the death cake. I awake at 21h00 to load up and start our push out of the AO. The drive was long and tiring. We coiled up out in the desert close to the LSA and bedded down for rest at 06h30. I was awoken at 07h30 for radio watch and was relieved an hour later. 09h30, the rest of the Marines wake up, and we head back to the LSA.

October 31 – Halloween, today is my little brother's 21st birthday. I feel that being out here ensuring Danny has a safe birthday is a good enough present. I do wish I were there with him celebrating though. At 10h45, we push to the LSA to refuel for our trip back to Payne. 13h00, we push out to coil up and rest some more before our long movement back to base. The lockdown on the phones was lifted, and I got a chance to call Danny and wish him a Happy Birthday. 16h00, the Chaplain came to our coil to provide religious service. At 19h00 we move back to the LSA to help provide security. We established a perimeter around Highlander forward with Alpha Company. HQ Bravo sets a small coil inside the perimeter, and we share conversations of our battles. Highlander blasted their 25mm cannons at the sky to celebrate mission completion. The Marines and I all cry out with victorious joy and the fact that we had sustained minimal casualties. I set up my sleeping bag and bed down for the night. 23h00 I was woken up for radio watch where I froze my ass off for an hour then went back to sleep after being relieved by CPL King.

Chapter 12: Mission Accomplished

November 1 – 05h30 reveille. We pack up and prepare to move out. While waiting to push, two CH53 E helicopters arrive and land right next to us to pick up the British Special Forces members that served alongside us on this mission. At 07h30, we start our movement back to Payne. Along the way, we were doing several recovery operations to downed or stuck vehicles. I felt very happy today for several reasons. This month I was expecting to be back in the United States enjoying the better qualities of life; also today ends the operations that will be doing in this country, or so I thought. I just wish time would tick a little faster so I can get the hell out of here already. After countless recovery ops, we finally made it back to Payne. Here is where we will stay until our flight home.

November 2 – We do vehicle maintenance all day and take everything out, and prep them for turnover to 3RD LAR. I ensured all extra medical gear used for our last operation was turned in to the BAS.

November 3 – Today was pretty much a slow day. There is nothing exciting to talk about today. We moved into the overflow tents that the Marines nicknamed District 9 because living conditions were a little shittier than our previous tents.

November 4 – Today, the Command General paid a visit to us and spread his congratulatory word of a job well done. We were proud. All-day after that, I spent it in the computer center getting all my Marines to do their PDHA (Post-Deployment-Health-Assessment).

November 5 – Nothing significant about today. Just another day closer to getting home. We gathered all of the Corpsmen in LAR together for a job well-done photoshoot.

November 6 – I went to the gym and sat around all day doing nothing except waiting to go home.

November 7 – I did nothing all morning. I got word we were going out again tomorrow to interdict west of Deeshew for 5 days. Everyone was pissed off that we were going out on another operation. Everyone just wants to relax and go the fuck home.

November 8 – We pack up and throw our extra gear in the 20-footer storage box that was to be shipped back to the states and prep the vehicles for a 5-day operation. 11h45 we go red count 1 and begin our movement out into the desert. We drove for a couple of hours after stopping at South Station to pick some Afghan Army members. We arrived at the outskirts of the village of Landary and linked up with Red Platoon at 14h20. We are here because there were reports of rocket attacks during the elections. I held rear security on Black-8 while our scout team punched a patrol to cordon and knock (kick down doors) at the target complex in the village. We were parked next to Black-8 in a wadi (dried river bed) depression surrounded by sand berms and green shrubs. As I scanned our 6, I saw large sand dunes in the distance and larger mountains in the further distance. 16h00, the scouts intercepted personnel we had sought out to interdict but needed the camera to get photographic evidence, so Black-8 and Black-7 punch out to the middle of the desert to link up with Black-4 and 5 to bring them and the camera back to the scouts. 17h30 We push to the center of the coil in the center of the village. We will arrive at 17h40. Red scouts walk to the back of my pig with several large bags. They unload into my pig saying opium. Each bag of raw opium total 5 bags the size of basketballs, weighed about 15lbs. The scouts also said they found several 157mm rockets within the compound as well. I gave a closer look at these bags and picked one up. The smell resembled that of a very potent chemical. It stunk up the entire vehicle. It looked and felt like hard black mud. The sun goes down, and we await our next order. 17h30, we slowly make our way back to South Station, escorting the Afghan Army members. We make it safely there, drop off Afghani members, then head back to the HQ coil outside of the village. We enter the coil at 21h00 and set up for a night's rest. I closed up the back of my pig because the temperature dropped drastically on the way over here to what feels to be about 50 degrees now. The potent smell of raw opium in the bags on the floor next to me kept me awake for a while before I finally fell asleep.

November 9 – Midnight, I was awoken to stand 2 hours of radio watch. It was balls cold, but luckily for me, Devin Brown just got off roving security post, jumped in the vehicle, and threw the heater on for me, and kept me company for the rest of my post. 02h00 I was relieved from watch and went back to sleep. 06h00 reveille was sounded. We packed up and prepared to travel deep into the main section of the village to talk with the village elder. 08h00, we go red count 1 and push into the village, passing several mud and stick made huts and farm fields. The people of the village went about their normal business, tending to their livestock and crops as we kicked up a dust cloud through. We stopped 2 men on a motorcycle to question them, asking where this village elder was located. They gladly agreed to lead us to his compound. We continued to push deeper into the village passing more farmland, shepherds herding cattle, farmers harvesting crops, mud huts, and trees sectioned off by small water irrigation creeks. After driving for about 10 minutes, we stopped outside a complex with ruins of an old tall mud castle next to it surrounded by large and wide crop fields. The scouts and interpreter punched out and approached the compound and made contact with the elder. Black-8 and 7 post-security about a football field away, and we observe. It was about 09h15, and all seemed to be peaceful about the land. About 1 hour goes by. Red scouts report locating a marijuana field not yet harvested 50-feet by 400 feet. Then 15 minutes later, they report discovering another marijuana field 200 by 200 foot around 1 month from harvest. 10h30, the scouts finish talking with the elder load up on the vehicles Black-6 and Black-4, and we head back to the HQ coil outside the village. We sit around in the coil doing nothing until 13h30. The CO decides to move to higher ground, so we load up and travel further west between the giant sand hills I was staring at for security the day we found all that opium which is still stinking up the back of this vehicle. 14h00 after we coil up overlooking part of the river, Red Platoon reports capturing a source that agrees to point out all the Taliban weapons caches in the valley. At 14h30 2 white pickup trucks filled with women, children, and farmer men drive near our coil. We search them and let them pass. 15h00 Red Platoon rogers up and reports a downed LAV not starting. 15h30, we load up and drive through large sand dunes for about 30 minutes with Red Platoon. The mechanics try to troubleshoot and fix the starter on one of red's vehicles. I hop out of Black-8 to refuel it. While waiting for the pump on the 7-ton refueler we took out here, I got the scoop on what's going on in this area by talking to Baran and Brannon of Red Platoon. They say they found 2 weapons caches and IED making material in the village they've been patrolling in. The mechanics couldn't figure out how to fix the problem, so we rigged the vehicle for tow behind the 7-ton. Night fell, and we pushed to Payne. The drive took several hours, but we made it back safely, dropped off the broken LAV, picked up EOD, and escorted them back to Red's position. Again, the drive took another several hours, and we almost went rolling down the side of a mountain off a cliff which was scary as shit because we were literally 3 feet from doom. Luckily, we saw that there was no ground beneath us, and 1st Sgt had the driver slam the breaks just in time. We made it back safely and parked in the center of the coil. By the time it took for me to check on a sick Marine, notify 1st Sgt my recommendation to pull him from the field, set up my sleeping bag, and crawl in for rest, it was midnight, and I fell asleep as soon as my head touched the helmet I used as a pillow.

November 10 – USMC Birthday. Reveille at 06h00. 08h00 EOD does a patrol out to the IED site and demolishes the pressure plate IED that was set in the mountains behind us. 08h45, they patrol back and mount up, and Red Platoon escorts them back to Payne. 09h00 HQ platoon mounts up and also heads back to Payne. It seems our Operation Last Show of Force has come to an end. The drive back took about 3 hours. As soon as we got to Payne, we cleaned out the pigs and settled back into the tents behind the BAS. Chow this evening was very good. There were decorations in the chow hall and monuments of the flag-raising on Mt. Suribachi. They served steak and lobster at the chow hall and had a large cake. Several hours after chow, I decided to go down to the gym. The place was packed with Marines. I guess everyone is trying to hurry up and get their swell on before going home; also, Marines from 3rd LAR are here to replace us, so everything is packed. I stayed for 20 minutes before getting frustrated that everything was being used, so I finished my workout with 100 pushups back at my tent then went to bed.

November 11 – **13h00** Nothing too spectacular today, just OFP (own fucking program) for me and turning in the remainder of my medical gear. White Platoon was still out in the field, interdicting and showing the presence of US forces still operating in the area.

November 12 – Nothing too exciting today, just MWR and gym all day.

November 13 – Another day of un-excitement. The Marines got their heights and weights taken. I talked to LT Naga on the phone because I had to call over to castle to find out about the rest of my Weapons platoon getting their PDHA's done over there. It was good to hear his voice again after several months. He and I got along great because we had a lot in common, and we understood each other well. He was a great medical officer with a lot of knowledge, and I loved learning from him.

November 14 – The temperature was getting colder. Even with a jacket and gloves, I would shiver at night and on my way to morning chow. Several elements from each platoon went out to do left seat, right seat with 3rd LAR to show them the AO today. Left seat right seat consisted of half of the members from our battalion and half of the members from our relief battalion going on orientation to the area missions outside of the wire. It was great to know they were finally here. That was a reassuring factor that we would be getting out of this hell hole soon. All morning CH-53 E Helicopters landed here, dropping off members from 3RD LAR and picking up members from 1st LAR Echo Company to begin their journey home. I went to the gym at 11h00 and just hung out all day. The rest of Weapons platoon arrived here from Shabu. Nothing much too exciting went on for the rest of the day after that. Every public facility was packed with Marines. There was a line that wrapped around the chow hall, the showers were always full, and the computer center had over an hour wait time. We were due to leave here on the 21st, so the anxiety and anticipation were eating everyone alive.

November 15 – I handed in my M4 rounds today. It felt so good to get rid of some extra weight—just the normal routine of waiting around again.

November 16 – Those several elements left again to do the Rip with 3rd LAR for 3 days. I had to make sure all of my shit and myself was always squared away because 1st Sgt and the rest of the staff NCOs were looking for things to do, and when they got bored, they started acting like assholes. It almost seemed as if it was pleasurable to the higher-ups to mess with the younger Marines just to keep themselves entertained. If they find one little thing wrong with you or your gear, they chew you out because it's fun for them when they have nothing better to do. I spent the whole morning cleaning my area and cleaning my M4. I also did like what other Marines here were doing by dumping out our packs and taking them to the showers to rinse the dust and dirt off. Customs was strict about letting dirty gear on the plane. I also rinsed off my flak jacket and helmet. I laid the gear out in the sun for several hours to dry and then packed all my gear back in the packs. I went to chow then to the gym. Everyone went about their business through the day. I played and finished a game of Monopoly with Orta, King, and Dony. Dony won, but I was his last opponent. Soon after that, I took medications because I was starting to get a sinus infection, and then I went to bed.

November 17 – The usual routine today. My sinus infection was getting worse, but I was managing with self-medication. I slept most of the day off to feel better and help the time pass.

November 18 – The rest of our company returned from doing the RIP (relief in place) with 3rd today. The computer center wasn't as packed as usual because Charlie Company left to go home late this evening. Marines from our unit were starting their journey home, and the anticipation was building as we watched them wave the middle finger at this place while leaving us temporarily behind.

November 19 – Everyone made sure all their gear was clean, and all their ammo rounds were turned in. This afternoon at 16h00, my Company had an award ceremony for my junior Corpsmen that earned their FMF pins during this deployment. We also handed all the Afghan Local Interpreters certificates of appreciation for working with us. After all the rewards were handed out, the company CO took some time to reflect on our deployment, saying that it was a miracle that most of us made it out of here alive. Later on, I went to the gym with Tien then went to bed.

November 20 – 08h00 Sgt Abelar woke me up to go to the gym. On the way, he told me that Donald Mckoy had died last night in a car accident back home. He was one of our Marines who made it all the way through our last Iraq deployment, in addition to one deployment before. As soon as he returned home, he was struck by a motor vehicle. Everyone was extremely sad about the news; he was with us in spirit in Afghanistan. He was a damn good Marine and will forever be in our hearts. After the gym, I went to chow, then came back to my tent and did some last-minute packing for our trip home tomorrow. 13h30 Doc Donny came into my tent to inform me about the 14h00 Master Chief's call at the chow hall. There were Sailors on base at the chow hall, including Seabees, to listen to a motivational speech from the Command Master Chief. That lasted several hours then I went about my business for the rest of the day. Everyone was excited to be leaving tomorrow. The Marines celebrated by clearing a large space in the middle of the tent and ground fighting each other for several hours.

November 21 – Today was a day of excitement but ended with a dramatic event. We all woke up at 07h00, went to chow, and then we gathered all our packs and staged them in a neat fashion in rows on the flight line next to the helicopter pad. We waited for several hours on the flight line, ensuring we had all of our gear and EDL. 11h50, the MV-22B Ospreys were due to arrive and transport all of us in sticks of 12 each to Camp Dwyre. I was amongst the first to go along with 11 other Marines who gathered their gear, lined up their packs, 3 each, and donned their flak and Kevlar and stood in line in alphabetical order waiting for the bird to land. It was so exciting hearing and seeing the first couple of Ospreys land several yards in front of us and start our journey home. The birds landed right on schedule. We loaded up with our packs on board and sat in the seats facing each other with smiles of excitement and tears of joy. It was my first Osprey ride, which was similar to that of riding on a helicopter. The only difference was more shaking from the dual rotating propellers moving us fast through the sky. I broke out my camera and took some snapshots and videos of the flight, but they are now lost somewhere in just my mind's memory archive. The transport to Dwyre took about 10 minutes. We safely landed on the flight pad and unloaded all our packs, and headed over to the 7-ton truck and bus waiting to transport us. After getting accountability of all gear and personnel, we loaded all packs on the 7-ton and loaded and crammed them onto the bus. We took a 10-minute ride to our tents on the other side of the base. It felt really strange riding on a bus because it had been 7 months since I'd been in one or do anything else but ride in an LAV in the desert. We unloaded from the bus and got settled into our new living space in the large insulated tent. After getting situated, we still had time to grab some excellent chow in the DFAC chow hall close by. The gym was much bigger here also. I broke out my Navy PT gear and went for a run on the treadmill, and lifted some weights. It felt good to run after 7 months of not doing it except for the times I ran while tacked up in combat situations. Quality of life was much better here compared to where we just came from. Shortly after the gym, I went back to the tent to change over back into my frog suit to go to evening chow. It was around 18h00, and I was rounding up some Marines to accompany me to the chow hall when one of my junior Corpsmen came to me to get some advice for an injured Marine. I stepped outside the tent to see LCPL Bonhill sitting there with a large bloodstain on the side of his blouse. I lifted his blouse to see a 1 inch long, ½ inch deep laceration on his left-right rib cage. It was a clean slice and looked like a stab wound. I decided to take him to the nearest medical facility for stitches. I notified my SGT about what was going on and took the injured Marine to a transport vehicle to get him to a medical facility. On the way, he told me he was stabbed by another Marine. I told my junior Corpsman to stay with him, and I returned to the tent to notify his platoon SGT of the news. There was an uproar, but I had to write a statement and report to higher. I went to chow then returned to question Bonhill during his treatment. He took 9 stitches, and he was placed on light, limited duty (LLD). Everyone in HQ platoon in Bravo Company had to pay for this event by playing stupid games such as marching to chow as a unit and reporting in every hour. This was done in order to avoid having to conduct paperwork and create an investigation which would have halted the process of us returning home. I did my hygiene thing and went to sleep on my cot. I was excluded from the extra instruction that the Marines were receiving for the incident. I am not sure what prompted this Marine to stab another. All I knew was to do my job, keep my mouth shut and nose out of it and back off into the shadows from the nonsense.

November 22 – Today was another exciting day. Reveille was sounded at 06h00. It was freezing outside of my sleeping bag. We did our hygiene and formed it up outside the tent after checking our serialized gear. As part of the punishment from yesterday's incident, 1st Sgt ordered us to march to chow as a platoon along with other things. I did not mind though. It instilled better discipline for garrison and reminded me I am still a part of the military. Chow here was really good. We marched back to our tent and waited for our Combat Transition class from the Chaplain. The class consisted of instruction in preparation of what it may mentally be like for us once we return home. "Don't stab your wife, don't kick your dog, don't scream at the kids, and call the suicide hotline in case of emergency from PTSD." At around 08h30, the RAS (Regimental Aid Station) Leading Petty Officer and the psych tech 1st class Petty Officer came in my tent and informed me that the Command Master Chief wanted to meet me and my

Corpsmen and talk with us. At this point, I felt honored and excited. I was ordered to have myself and my junior Corpsmen arrive at the Regimental Combat Team 1 RAS at 09h00. I rounded up junior Sailors and notified their sergeants where we were going, and started walking down the road. HM1 saw us walking and stopped the bus he was riding on so we could hitch a ride. We drove about half of a mile and got dropped off at our destination right on time. We were led into a nearby classroom where a good handful of other 1st class Corpsmen showed up and greeted us as their honored guests. I guess they were there to meet the Master Chief as well. We waited until about 10h30. The Master Chief walked in, and everyone stood upon his arrival to show our respect. Master Chief said to my Corpsmen and I, "So, you're the men I've heard about. Please have a seat. I come down here today because I wanted to personally thank you, men, for what you have done out there. You're the reason we get so much praise and recognition for the work we do. You all stepped up to the plate and did what America asked you to do out there. I am here to thank you and tell you that you have done a great job, gents!" He presented his special coin to each of us and gave them to us with a handshake and hug, and said, "Have a safe trip home, men; it is your time to celebrate and enjoy yourselves with loved ones now." I was so proud of my men and myself at this special event. I am always proud. I was so grateful to receive recognition for our hard work here. As we left the tent so Master Chief could start his meetings with the 1st classes in there, everyone gave us head nods and smiles as we passed them. I was so honored I found myself fighting to hold back the tears of joy. We were about to start our journey back to our living tent when the Leading Chief Petty Officer of the Regimental Combat Team stepped outside to talk with us. His name was Chief Pierson, and he gave us one of the most motivational speeches I have ever heard from anyone. My favorite part of his speech was, "Men, I am proud of you all standing before me, you did the right things out there, and you heroes are the reason why I still, after 20 years of service, can go to any bar wearing a Corpsman shirt and get a free drink, hugs, and praise from veterans. Don't let anyone ever tell you that you didn't do the right thing out there. You did what America asked you to do, and you were there for your Marines, and by God, you saved everyone's lives and brought them all home. Good job, men, have a drink for me when you get back!" He shook our hands, and we were on our way. We made it back in time to still catch the Chaplain's Combat Transition Class on dealing with combat stress and starting that integration back into the world left seven months ago. Immediately after the class was over, I formed it up with my platoon, and we marched to noon chow. Upon returning from chow, everyone took a nap for several hours because we were tired. A majority of the Marines were sick with head colds due to the weather change, and they were drowsy from the medication given by yours truly. We slept until 18h00 and went to evening chow together. At 20h00 I went to the gym with SGT H. for a couple of hours, showered off, and then went about my business until I fell asleep.

November 23 – It was just the normal routine today, marching to chow and working out at the gym. More than half of my Marines were sick with the cold. I did my sick call rounds this evening, medicating everyone. I notified the 1st Sergeant about all our Marines' statuses, and he came in to make fun of everyone with glee. I found this hilarious. I, too, was starting to get the sniffles, but I was able to manage. I also passed out shots of Nyquil and, within the hour, watched my Marines pass out to sleep like they were kids of my own.

November 24 – 06h30 reveille. Those Marines who I recommended to stay in bed stayed in bed, and the rest of us marched to chow. After chow, I slept in all morning until noon chow. The Marines played football on the gravel this afternoon. I finally got a chance to go to the MWR after they lifted the punishment for the stabbing incident. We no longer had to march to chow and were given back the freedom to go where we wanted as long as we had a battle buddy and, of course, don't stab one another again. I went to chow with the HQ NCO's then to the gym, and just hung out praying for time to go faster.

November 25 – The Marines awoke at 06h30 to hygiene, shave and check serialized gear. Everyone wasn't too excited about it being Thanksgiving because we were still here in Afghanistan, but we made the best of it today. We followed the American tradition of watching football and overeating good food. This afternoon the entire company gathered on a dusty old construction field to have a football tournament. Each platoon played against each other for several hours. During the last game, 1st Sergeant got slammed on his eye from a collision with another Marine to catch a pass. "Corpsman Up!" 1st Sergeant walked over to me with blood all over his right eye. I grabbed a water bottle and took off my shirt to wipe the blood and irrigate his eye. He had a nice ½ inch slice, Rocky Balboa style, just under his eyebrow. Assuming he would need stitches, I brought him over to the RAS to clean it out. I sat him down in one of the treatment rooms and started scrubbing his cut with an alcohol swab. It turns out the slice wasn't so bad, so I closed it up with a butterfly band-aid and sent him on his way to finish the game. Blue platoon ended up winning the tournament for bragging rights. We police called all the trash from the field and hit the showers. We changed over to our cammies we would normally wear in garrison, and Devin Brown, myself, and several other Marines made our way to Thanksgiving dinner. We all sat down together in the Thanksgiving decorated chow hall, said grace and ate some good chow. It was nothing compared to Mom's delicious cooking back home, but I made the best of it. After our return from chow, we received word that we were getting on the plane to head to our next destination in the country of Kyrgyzstan, a base location further west in central Asia. Like here in Dwyre, it was one of our destinations 7 months ago on our way out here. I remember it like it was last week. The feelings of going there were different because the hell had been walked through, and the deployment was coming to an end. There was one thing at this point that I was thankful for about this deployment, and that was that it went by fairly fast. It was much faster than my Iraq deployment. As soon as word was passed, we all excitingly packed our gear, did gear and ID card checks, and staggered our gear outside the tents in pack designation order, and continued to go about our business anxiously awaiting tomorrow morning's trip.

November 26 – Reveille was sounded at 04h30. We packed our things and stood outside the tent for several hours in the freezing cold, waiting for the shuttle bus to take us to the flight deck. The bus came, picked us up 12 at a time, and shuttled us to a large tent next to the landing strip. Everyone huddled around on the gravel floor in the tent, trying to stay warm while we waited for our flight. At 09h00, the C-17 arrived, and we all gathered and lined up to get on. The plane ride took about 45 minutes, and we safely landed in Manas, Kyrgyzstan. This was a U.S. Air Base that was used to transport troops to and from the Middle East. It was operated by the U.S. Air Force, so the quality of life here was even better than the base we just came from. Troops who were stationed here were allowed two beers per night, had access to amazing chow halls, a gym, a movie theater, and even a bowling alley. It was at this point where several of my Marines were making a comment "Well, I sure joined the wrong fucking branch of the military." We were no longer considered to be in a combat zone at this point of our journey. The weather here reminded me of upstate NY, cold, but the air was fresh. We all gathered on Greyhound-style buses and took a ride through this beautiful Air Force base to our living quarters. The tents had heat and mattress bunk beds. We got situated with linen and went to chow. The Air Force had it so nice here. It was good to walk around and not have to carry my M4, yet it felt so strange everywhere I would go. The mental transition out of the combat mindset was beginning to affect us. There was a strange silence that we would all express as if we weren't nearly ready to decompress from the stressors that we had become accustomed to. I was so excited to have all these nice things again, however, I ended up staying awake the entire night using my own laptop and talking with loved ones on video chat. The time zone was different here as well. We were 14 hours ahead of California. I wanted to adapt to that, so I wasn't tired all day, so I slept the entire next day through the 27th from 08h00 until 18h30.

November 27 – 19h00, we went to US customs with our luggage so they could inspect and make sure we weren't smuggling anything illegal back into the states. That took about several hours and wasn't nearly as painful as when we had to go through customs on the way home from Iraq. I stayed up all night again hanging out in the Morale and Welfare Center.

November 28 – I went to bed at 08h30 and slept until 18h00. I took a shower, then went to chow, where they served steak and lobster, then just relaxed, waiting to go back through customs with my carry-on luggage and get on the jumbo jet to fly back to America. 23h00, everyone heads out to the customs tent to stand in a massive line and wait to be searched. About an hour after, I get through and get my carry-on bag and laptop cases searched, then sit in the waiting tent for several hours.

November 29 – 01h30, everyone made it through customs, and we gathered outside in a mass formation of over 280 service members and were herded like cattle over to the terminal. This is where we sat for several more hours watching football on the projector and waiting for the working party to finish loading our luggage in the belly of the jumbo jet. 05h30 – We all loaded up on the buses that drove us to the landing strip. As we approached the flight deck, we could see the two-story jumbo jet prepped and ready to take us home. It was a sight for sore eyes this time around. We unloaded from the busses and waited at the stairs of the plane in the freezing cold. After a painfully freezing 25 minutes, we started walking up the stairs and entered the bird. I sat next to 2 Army guys and fell asleep as the plane lifted into the sky. We arrived in Anchorage, Alaska for our layover. I could see snow on the ground and felt a freezing cold breeze of air that flowed into the plane as the doors opened. I took in a deep breath and exhaled, saying "Ahh, freedom." We offloaded into the terminal. Everything was closed, and the stores were caged up, so there was not much to do but stretch my legs and walk around in the hallways like a caged animal. I found a power outlet in the wall, sat down, and started texting and calling loved ones, letting them know that I was back in the states. It felt good to use my cell phone again. After about an hour and a half of hanging out, we loaded back up on the plane and prepared for the take-off. Next stop, March Air Force Base California and then busses to Camp Pendleton.

We landed in California while it was still the early afternoon and loaded on the buses that would take us back to Camp Pendleton. I can remember looking back at some of the Marines seeing smiles, anxiety, tears rolling down their faces, and middle fingers. There was a busload of mixed emotions going through all of us at this point. To our surprise, there was a convoy of veterans on motorcycles, hundreds of them revving their motors, blaring horns, clapping, cheering, and giving us head nods and thumbs up. On the flight deck, there were also many citizens lined up cheering and waving American flags on either side of us as we transferred like a parade from the plane to the buses. The buses started, and we began rolling back to base. As we took off, we were escorted by the veterans on motorcycles. They stopped traffic for the buses we were on, led and followed behind, escorting us all the way back to our home base. It was so badass to witness what they did for us, unlike what Viet Nam Veterans experienced when they returned home. I and many others cried tears of pride and joy several times as we were honored for our struggle. We finally reached base later that evening, where there was yet another crowd of citizens waiting to welcome us in a big cheer. This time, it consisted of our friends and family members. As we reached the ramp where we would stage the vehicles, some of the Marines were crying after spotting their loved ones. Some were shouting out loud, "Get me off this fucking bus!" Some Marines did not have loved ones specifically there waiting for them as we arrived, so their excitement wasn't as intense as others. We all unloaded from the buses. We were called to the last deployment formation. The order was given "Now is time to spend with loved ones, job well-done men. DISMISSED!" The Afghanistan deployment was finally officially over! We all let out a large cheer and ran to our loved ones jumping into each other's arms. This was really it. It was finally over. The frustration, the suffering, the hardships, the sleepless nights, the fighting, the mental and physical pain, and strain we were finally free.

Or were we...

Chapter 13: The Aftershock

After spending the weekend with our loved ones on liberty, we returned to base. Some were assigned to barracks rooms and were preparing for leave to travel to their homes throughout the country. I remember trying to go out in public on my own, feeling nervous, panicked, anxious, and severely distant from the rest of the world. I was having nightmares of still being in Afghanistan every single night. I would wake up in a panic and covered in sweat and terror, looking for my rifle and gear. Every time I opened my eyes from the sleep that was very difficult to obtain, it would take me several minutes to realize I was no longer over there. I was placed in a barracks room by myself and kept to myself for the most part. I was trying to decipher what was going on with me mentally. The adjustment phase from what I had just gone through became more difficult as the days went on. I planned my trip home on post-deployment leave. This was where I was getting excited to finally see my loved ones.

I made my journey from California to upstate NY. Upon walking out of the terminal, I could see my father standing there waiting by the car with a very big smile on his face wearing his Navy Veteran hat. I swiftly walked up to him, and he pleasantly greeted me, "Welcome home, Sailor!" I cried as I gave him a very big hug. He looked into my face and asked if I was okay. "I'm fine, Pop, let's go home," I replied as I wiped the tears.

The car ride was about an hour, and not a single word was said about what I went through. Dad was catching me up on current events and talking with me about guns to keep my nerves calm. I didn't speak much, just listened. He had a lot to talk about, and I would humbly listen to some of his rants. It was as if I had never left home. He was still his good ole' self. We reached the house where I grew up, and mom was there with tears in her eyes, smiling and with open arms. I could tell they were both concerned. They wanted to ask me many questions, and I could tell they were monitoring me closely. I would tell them I was fine and did my best to act as if nothing was bothering me. Seeing my little brother Danny and my very close friends, the Beairstos, helped me come back to this reality at times as well. We would goof around and enjoy outdoor activities together. I was eager to deer hunt.

Time went on, and we interacted with each other as a family just as we did before I left for deployment. I could tell that both my mother and father could sense something was different about me. They could tell I was trying to hide emotions and anguish. Mom was always a night owl, so I'm sure one of the nights while I was home, she may have heard me screaming in my sleep or waking up in a panic. I was distant in public. I was very quiet and always looking around on guard. Dad found it strange, and it got to the point where he was fed up with being quiet about it. He noticed me out on the bank, standing near the river staring out into space on several occasions. This was abnormal for me because I'd usually always have a fishing pole. One day while I was still home on leave, I was standing at the riverbank in front of the house again. Dad approached from our house behind me. He noticed I had been crying. He didn't ask what was wrong because he knew. He put his arm around me and said, "Nicky, we need to get you some help, Shipmate." I tried to tell him I'll be fine and honestly thought I would after some more time.

The next day, he took me to one of his friend's veteran service offices, where they offered help in any way they could. I learned about the different services that they provide for all veterans. It had a huge impact on me that day.

It soon became time for leave to end, and headed back to base in California. I said my goodbyes to all my friends and family. I could tell they were relieved that I was home safe, yet even more concerned because I was a different man. I reached my destination back to base. Waiting for me was a girl who I had been talking with throughout deployment from the local area. We had developed a relationship and became serious with each other. We spent as many days as we could together, and she helped me escape negative emotions a lot. As time went on, it was getting closer for me to pick orders to a new duty station. I was torn between the decision of staying in the area and ending the relationship with her or finally being stationed somewhere close to where I had my family and friends and where I could get the help I really needed. I wanted to have both a relationship with this girl and be close to home. I received orders to Newport, Rhode Island.

She was sad about the fact that I would be on the other side of the country for several years at least. I was pressured with the decision of wanting to keep her in my life, and time was running out. I decided to ask her to come with me as my bride. It was a typical young military marriage where it eventually didn't work out. We lived together for just over a year before things went severely sour. We were still young, and she ended up cheating on me. I needed to work on my issues from war that were preventing me from having a relationship with someone and functioning as a good husband. Not pinning the blame entirely on her for cheating on me, but I felt wronged, and after sending her away back to California, it gave me time to reflect on myself and the things I could do to improve my issues of being distant from everyone and everything. I would shut myself out from the world and not be able to think straight or function correctly. Being married takes a lot of time, concentration, attention to one another, and many other things. They were things I was struggling to do for even myself. I blame war. It cost me a marriage, but it was also a learning and developing experience. Time went on, and I was eventually able to focus more on positive things that would keep my mind occupied. I was still in the Navy at the time, but on the blue side of it versus the green with the Marines. I worked at the Naval Health Clinic and held several important positions as King Hall Medical Department's - Assistant Leading Petty Officer, Education and Training Department's - Tactical Combat Casualty Care Instructor, and then was recommended to one of the top positions in the command as an enlisted member working in the Command Suite as the Legal Clerk with Shane Marion, a Naval Officer Veteran and very dear friend and mentor.

    I found myself discovering new ways to try to cope with the combat stress and other internal and external injuries that I have sustained while on the green side with the Marines. Bodies and minds are not designed to endure the constant stress of tactical training and deployment. Working very hard and staying extremely busy from sunup till sundown. I kept my mind focused on the present or the future instead of being dragged backwards. I focused my energy by competing for and achieving Sailor of the Quarter. Then I was selected for Sailor of the Year, and finally, getting promoted to E-5 Petty Officer Second Class. I accomplished that along with a lot of fishing for striped bass, hunting for pheasant and deer, volunteering, earning an Associates and a Bachelor's degree, prescription medications, mental therapy sessions, and copious amounts of nicotine and masturbation. What I did not know was that all these things were masking the underlying pain. The pains, both physical and mental, were subsided for the time being because I was ignoring them. I was smothering them with other activities and excelling in life. I was on top of my game and thought that I had it all under control for the most part. It wasn't. I would still experience mental and physical pains. Nightmares, feeling and acting distant, crying for what seemed to be no reason, and feeling even more lonely than ever.

I was missing Molly tremendously and would remember how she would make me feel every instance where I got to spend time with her on deployment. Memories of joy with her made me decide to get a cute as a button, blue-eyed, chubby little chocolate lab puppy as my companion whom I named Toby. We instantly became best of friends and spent every day together. I trained him very well and would teach him similar commands that we would use with Molly overseas. Toby would help me cope with the pain on a substantial level. After maturing, he would even begin to be able to tell when I would encounter an anxiety attack, fall into a depressed mood, or not be able to focus. He would always be able to pull me out of the hole I would frequently fall back into on a daily basis with his love, affection, goofy personality, and extreme attentiveness.

After achieving the rank of E-5, I encountered an attractive blonde-haired Sailor who arrived at the command. As usual with everyone and everything besides the previously mentioned items that I used to blanket my pains, I kept my distance from her. After working together for several months, we discovered that we had some things in common. Still keeping my distance, she would give me quaint signals that she was into me. I eventually let my broken-hearted guard down and my physical and emotional attraction take over, and decided to believe that she would be worth another shot at love. She, Toby, and I spent a lot of time together. We really enjoyed each other's company and doing many different activities together. Coincidentally, our hometowns were only an hour apart from each other. She had portrayed that she also enjoyed all the hobbies that I did. Her mother and stepfather were also very loving, and we shared much in common and loved spending time with each other. We helped each other with moving forward with finalizing our divorces from our previous lovers and decided to tie the knot with each other shortly thereafter. I had mentioned to her and her family that I had many issues because of war and that it would be difficult at times to love me. It was true at times for her. She had also portrayed that she too had a difficult past with family and ex-husband, and we were both willing to help each other out and enjoy life together.

She had only a short time in the Navy, and it was coming close to the time where I had to make the decision to either stay in or get out. I knew that if I stayed in, a relationship would be even more difficult after I reenlisted. I decided instead to gamble on something different and to get out and give my life to this new woman. I spent a total of 8 years as a Sailor in the U.S Navy and had achieved so much, including the title of HM2 (FMF) Bertucci, USN. My future plans of becoming a Naval Officer were no longer in my sights.

My father immediately jumped on my case and instructed me to file for disability compensation with the Department of Veterans Affairs (VA). "Contact your Veteran Service Officer (VSO) right now Nick" he said with deep concern and sincerity. At first, I was reluctant and felt undeserving. It is incredibly difficult to admit a form of weakness or pain. Still, I finally filed my claim and was rated for my issues. My spouse eventually received orders to Beaufort Naval Hospital in South Carolina. We lived there for several years, where I obtained a master's in wildlife management. The pains from war were beginning to emerge again because life had been slowing down, compared to being in the Navy. I would do my best to stay active and productive in order to drown out the pains. My neighbor at the time, "Uncle Rick," a South Carolina native whom I had become very close with, had always commented on the fact that I would never slow down and continue work from the moment I woke up till the moment I went to bed. My spouse was becoming miserable. She would always complain that she hated being in the Navy and grew tired of it because the command and morale were dismal. I could no longer relate to her on this aspect because I was no longer in, and we had vastly different experiences in our separate Naval careers. What

I wasn't realizing was that she was also becoming miserable because she did not actually enjoy hunting, fishing, guns, and many other outdoor things, and my problems from war were slowly starting to get the better of me again. I was struggling to put on what many would call "The face" that would mask the underlying physical and mental pains of combat stress. We were still very much in love and supportive of one another because we had promised a commitment to each other, so we pressed on to a new chapter in life.

Chapter 14:
A New Life Mission

It became time for her to end her enlistment. Without a second thought, she made her decision to get out. She had the desire to travel around the country while living in a van. I was nervous about the idea. She stated that she was going on this trip with or without me. I gave up a very enjoyable outdoor job that I had worked so hard at for the sake of her happiness and mine and to keep her by my side. We decided to sell everything we owned and built a van that we could live out of and traveled all around the country making adventure videos. It was a spectacular trip, and Toby and I really enjoyed it. I had my issues from war as usual, and they were ramping up worse than ever. I panicked and did not want this trip to end because I had seen how happy it made my wife. I decided to mask and try to ignore how rolling around in a tightly compacted vehicle, living out of it, visiting unchartered territory, and always being on guard brought back horrible flashbacks of living out of the LAVs during deployment. I had nightmares almost every night but would pretend as if everything were fine. It was hard to focus, but I did my absolute best to try to enjoy and have fun. We went from NY to Florida and made a few loops through the country and all the way to California and up to Washington State. I took the opportunity to reach out to and visit as many of my former Marines that had dispersed as I could throughout the travels. Forever affectionately being "Doc" in these men's eyes and still having the desire to care for them, I would mentor and urge them to seek help from a VSO, get seen by the VA, and file a claim for their military issues. It was so nice seeing the ones that I was able to again. It had been almost 8 years since our deployment together. It was like we hadn't skipped a beat even though it had been almost a decade. I remember being woken by my wife from a violent nightmare because I was convulsing, shaking, and screaming in my sleep. I remember dreaming about the explosions and hearing the machine gun fire, people screaming in battle, and the day I killed two people with CPL King and Loui O using the M240 machine gun. These men with wide smiles and quick with big bear hugs, when sitting still for a moment, they knew. They were right there with me, sitting at once next to each other and both with the other foot in the sand, lost in a place far away.

Toby was about 8 years old and beginning to grow tired from all the hiking when we decided to end our van trip. We made our journey back to NY, where the three of us spent the winter of 2018 with our loved ones bouncing back and forth between each other's parents' houses. Life was now at a very slow pace, and it was the dead of winter in upstate NY. After building a new, improved van where we could still travel and not feel bad about leaving Toby behind in it while going on hikes, we were convinced by her parents to move out to the Midwest due to several different factors. Riots were breaking out, laws and policies were changing that would restrict our freedoms of guns, hunting, and taxes were atrocious. We sat around all winter being bored. It was another factor that reminded me of deployment and made the negativity from war emerge even stronger. I was ashamed of my private war and so I hid it as best I could.

Chapter 15:
Hell Comes Knocking

In the spring of 2019, we finally found a house in Arkansas. We had sold the new van we had just built without even going on a trip with it, packed our personal belongings from it, and moved into the new house. We worked all summer to make the house and property our home. I could not bear to hear the noises of other human beings around me and had no desire to interact closely with anyone except for my immediate family, my wife, and Toby, so we agreed that in the middle of the woods was the place for us. We were both very excited for this new chapter in life with each other and pleased that we had found our dream home. Real-life had set in. We were playing house like a normal married couple. Life was good for the most part. It was slow because we both weren't working yet, but I was staying busy with modifying the house. The house was done being modified by the time hunting season came around. My wife got a job working as a manager at a big corporation, and I landed a career working for the Arkansas Department of Veterans Affairs as a District Veterans Service Officer. For me, it was another dream come true. Real-life had set in even more at this point. Routine was becoming a daily thing, and life was slowing back down. The issues from war arose again, but they were controlled because my wife and Toby were supportive and always by my side. As time went on, I could notice that something wasn't well with her. She had started becoming miserable again, complaining about life and drinking more wine than usual. I had promised her that everything would be alright and that we would do our van travels again. I was getting worried that she may have also been getting fed up with the issues that I still had from war. As the winter approached, she was becoming distant, less intimate, and less communicative. At this point, I opened up to her about my issues again and told her I was seeking help from the VA. It was too late. Just like how she did with me to her first ex back in our Navy days, she had found another person from her work that she moved on to. Very abruptly, the night before Thanksgiving, she told me she was leaving. I soon after found out that she had been doing things with this other person from work behind my back for quite some time. I was devastated and gave every effort to fight to keep her. She had made up her mind and abandoned me. From this came a full blast of every single negative emotion and nightmare from not only my past marriage that had failed in a very similar way but from war as well. It was gates that had burst and flooded my entire body and mind with negativity. I couldn't breathe. My Post Traumatic Stress had risen to the highest strength that it had ever been at. I found myself waking up every day thinking something was terribly wrong with me, and war was to blame. I would wake up from nightmares every single night into panic attacks, suicidal ideations, vomiting, constantly crying, unable to breathe, unable to function at work, consistently feeling drained, not able to eat, and completely shutting myself off from the entire world again this time, including my own immediate family. I talked to Toby every day and held him close while crying as he struggled to keep me calm. I felt as if there was no light at the end of this tunnel of hell. I was drowning in my own emotions from every horrible event that occurred in my past since experiencing war. I felt as if everything was my fault because I didn't receive the help that I so desperately needed from the moment I got back from Afghanistan. I felt as all hope was lost, and I couldn't find the strength to even get out of bed. Poor Toby suffered with me. I'm sure seeing your best friend in so much pain wasn't pleasant for him in the least bit, but he never left my side. I prayed for answers and a way out. I prayed that God would take me from this pain in any way, even if it meant leaving Earth for good. It went on like this for months. I was worried that I would lose my job because I couldn't function. I finally summed up the courage to reach out to my parents. They and the rest of the family were devastated, confused, and shocked from the abrupt abandonment. They were even more worried because they knew what it was doing to me internally, considering I had also mentioned that war had deployed to me again and hell had kicked down my door. I was afraid to fall asleep. It was a nightmare that I thought I would never escape. I had never felt so alone.

Chapter 16: Heroes' Hands In Healing Time

My prayers had eventually been answered. The answers had been there all along, but I couldn't see or comprehend them at the time. The only thing keeping me going was the thought of how important my career was to me. The fact that I was a part of something that's sole purpose was to help veterans in need no matter what. My desire to aid and assist other veterans who struggled like me made me want to live to serve them, but the answer wasn't that; it was to let them now serve and help me, too. It was time for me to start reaching out to others for help. First off, I went to my boss Gina Chandler, the Assistant Director of the Veteran assistance organization that I currently work for, the Arkansas Department of Veterans Affairs. She is also a veteran, as are all veterans who are members of this organization by requirement. She became like a big sister to me and offered me comfort in support and as much time off as I needed in order to get through my struggles. I felt extremely lucky to be blessed with such an understanding supervisor and mentor. I kept pressing on from there. I started reaching out to many of my Marines through social media and telephone. They were supportive of me and helped me get through this nightmare as well. Clint Walshart from White Platoon even drove 2,200 miles all the way out from just north of Seattle, Washington, to stay with me for a few months. He was suffering from his war pains in a different way but in the same sense. For two months, we spent every day together, coaching and guiding each other. We had become closer than ever before until he created a life for himself and moved elsewhere. We owe each other our lives on so many levels. We all do to each other. Since this whole nightmare re-emerged, I began seeking mental care via video conference only. I was still having trouble venturing out from my cave in the woods without having "the face" on.

The VA tried putting me on psychotropic drugs to help ease the pains, but I was reluctant to take anything that would alter my mind and did not want to become dependent on anything unnatural. Slowly, I learned to become independent again. I built strength. I learned to deal with the loneliness and stay productive. Loving neighbors had become my saving grace because all of my immediate family lives very far away. I made more friends within the veteran community here where I live as well, and they too have helped guide and inspire me. Letting music take a positive grasp on me was important on this journey. It would help me deal with emotions that I oftentimes had trouble with. I decided to create my own business where I would take people on outdoor adventures and teach them about wildlife. This adventure business became well known, and it served others as an escape from the hustle and bustle of a busy, stressful life and helped other military veterans cope with stressors that they have endured while in the military. I named it *Bertucci's Country Cabin*.

I reached out to the Northeast Arkansas Wounded Warriors Program and went on outdoor excursions with many different veterans from the local area. Naturally, we would all become friends over time. I found something that was helping me become somewhat normal again. I would be able to relate to other combat veterans, knowing that they too have suffered similar struggles or were still in pain in one way or the other. I have had the time and resources to begin my journey of making it through this nightmare. I have discovered that the path to salvation from this was not through things that would smother or mask the pain. For many, it would be easy to allow something bad to control you as long as it smothered the pains from war. Everything from alcohol, drugs, porn, crime, pretending everything is okay, over productiveness, to shutting yourself out from the world. These are all things that have the potential to be a temporary fix but lead to self-destruction. Thanks to others reaching their hands down into the fire pit I had yet again sunk into, I found the courage to grab hold and start climbing back out.

Chapter 17: The Final Message

To all those who have also suffered from war or other military hardships, my message is this; it affects everyone differently. Everyone has a different version of their story to tell and different ways of coping with it. Some may still be struggling to cope. I still struggle every day. I have learned to accept that and keep fighting the war that wages on in my head. Heed my words and be motivated. Reach out to others. Seek the help you need, offer your love and tell others you love them often. Be brave. Swallow your pride. Learn to love yourself. Think ahead, and don't dwell on the terrors of the past, or they will continue to control you or keep you in the darkness. Your future is bright. There are those who are ready to stand and fight with you. You stepped up to the plate to serve others, even if that meant giving your life. It is now time to recognize that there are ways to get help. There are thousands of organizations and programs filled with people willing to reach out their hands and pull you back from pains that many will never understand. Allow others to serve you so that you too may now strive for a better life and, in turn, serve others. **You are worth it and deserve every effort that others put forth to get you on track toward a good and normal life. It is never too late.**

I struggle every day. Negativity attacks me without warning on a frequent basis. There is no distress beacon that flashes for others to see when I am in pain, but I can assure you that if there were, I'd have to change the burnt-out bulb on a weekly basis. It is very difficult to try to control how it has and still is changing me. I've struggled with suicidal ideation, paranoia, and agoraphobia. I find myself being distant from crowds, standing off to the side or away from everything in the corner. I have learned that they are just thoughts that keep coming because of what my pains and experiences have developed into. I often find myself depressed, feeling like an 85-year-old man trying to move around or even get out of bed and fight through the physical pain. I would wake up feeling even more tired than when I tried falling asleep the night before because of all the tossing and turning. The military has trained me to be a light sleeper, so any noise would wake me into a panic in search of my rifle and battle gear. The emotional pain on top of and because of that has been depressing to the point where I wonder, "What am I even still doing this life thing for?" What helps me personally is attending events that gather fellow veterans in outdoor activities such as fishing or hunting. We would team up with one or two other veterans or outdoor guides that would take us on outdoor adventures. This, to me is therapy. I would still stay in the corner or away from the crowds at these events. I've been placed on and have tried countless forms of group therapy and medications issued by the VA. None of it ever worked. The anxiety attacks that I have developed have frightened me to the point where I tried meds that I never thought I would. There have been times where I would try to take a vacation from work and wouldn't be able to enjoy myself fully because of it. Being in a new place other than tucked far away from society in the woods has always been very difficult.

I worked very hard in life, getting an education and a good job where I could afford to live in a place away from everyone and everything. I have been blessed enough to land a job where I can now work primarily from home due to my issues. I have been able to find things to help me live a life where I wouldn't have to be placed in environments and situations that would aggravate my combat stress. I forced myself through the pains and struggles in order to achieve goals that would accommodate my needs as a damaged man. When I would have thoughts of suicide or depression, I would go outside and do something productive. It is important to me that I accomplish something that I can be proud of every day in order to keep my mind occupied on positivity rather than negativity. It was always up to me to work hard to accomplish my dream of living away from everyone in the woods with complete privacy.

Naturally, I love caring for others and giving back to society. I have decided that I wanted to share this place I now live in and have made it beautiful and tranquil for others to escape to and enjoy as well. I have decided to create a cabin in the woods getaway where people can come to enjoy peace, connect with nature, enjoy the outdoors. I would interact with the folks who would book the cabin to let them know who I am and that I am here if they ever need anything. Veterans come out to get away from it all and enjoy the sounds of nature, a campfire, fishing or kayaking in the local river, ATV riding or hiking through the trails, and have no one around to bother them. I've named this place *Bertucci's Country Cabin*. People can book a stay of any duration through online rental platforms or directly by calling me after discovering my website ARcountrycabins.com. In one year's time, it has become one of the most desired travel and relaxation destinations in Northeast Arkansas. I have been able to provide amazing experiences for many people seeking tranquility outside of their normal life routine, and that makes me happy. It has kept me busy and motivated to keep taking care of others outside of work. Occupation outside of my own thoughts of negativity have been what I have found to be what keeps me going every day. I refuse to give in for the sake of being there for others who need me or need what I have created for myself to share.

Use this message as fuel for energy to reflect on how what you have been through has made you a better person and the rest of the world a better and safer place to live. Always be kind and humble to others in every situation. Even though you struggle with pains from the past, don't let those pains make who you are or make you become something ugly or unpleasant to the rest of the world. The way to win this war within is to become something better and be above that storm from the past. When negativity hits, I reflect on the positive aspects I have learned. Take your pain and use it as motivation to create something to escape it. It is okay to let others know you are in pain so that they are aware of your past. It is not okay to treat others poorly because of the scars you wear.

People who visit me look around and are very impressed with where I live because of how beautiful and creative it has become since I've moved here. They would ask, "How did you find this place?" I would reply, "I didn't; I created it. This place reflects my life. It was once a mess. A mess when I got here. I've been through a lot of pain and suffering in my life and have worked hard through it in order to make it what it is here and now." I have discovered an appreciation for the things that were once taken from me and that others in the world still don't have, and that many take for granted every day. This consisted of clean water, shelter, electricity, food, culture and diversity, loving families, good mentors, and positive resources filled with people who are proud of you for what you have done for this great nation no matter what. God Bless you, God bless those who are still fighting, God bless those who serve members who have served, and God bless this great Nation and all her allies. The war wages on and forever will.

Never give up.

Photo Gallery

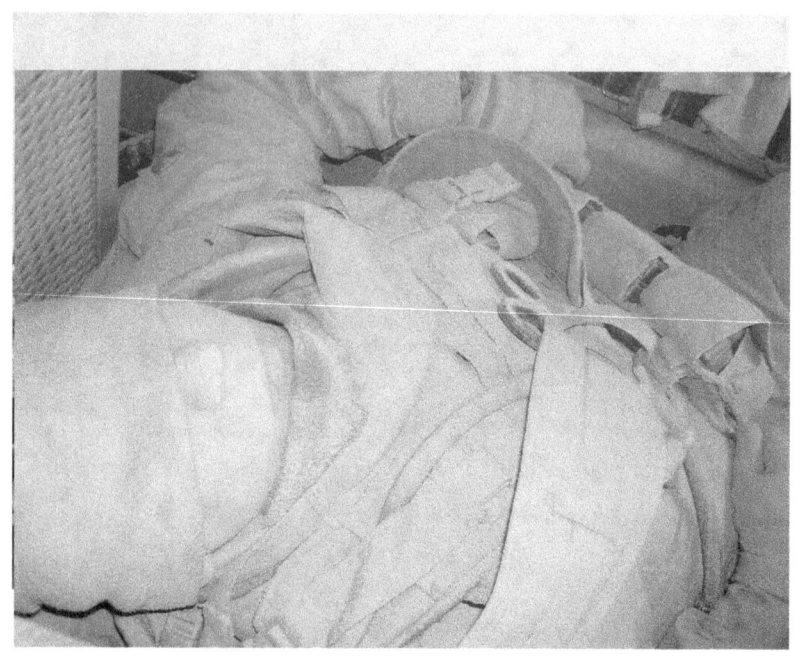

Nicholas Bertucci

Page 175: Awards: Fleet Marine Force Warfare Device, Combat Action, Navy Unit Commendation, Good Conduct with Bronze Star, National Defense Service, Afghanistan Campaign, Iraq Campaign, War on Terrorism Service, Outstanding Volunteer Service, National Sea Service Deployment, North Atlantic Treaty Organization, Expert Navy Rifle Marksmanship, Navy Pistol Marksmanship, Dog Tag, Name Tag, Rank and Rating and the different colors of uniform.

Page 176, top: Marines Minesweeping on a dismounted patrol on hilltops aside the Helmand River.

Page 176, bottom: A day to cool off in the Helmand River to boost morale while outside the wire. HM2 Bertucci, CPL Orta, LCPL Hernandez, Afghan Interpreter, and Bomb dog Molly.

Page 177, top: M2 Bertucci is kneeling with Afghani Truck Seizure of hundreds of pounds of raw black-tar opium and some smaller bags of processed opium. Detained 2 afghan locals attempting to transport across the desert in Helmand Province.

Page 177, bottom: Light Armored Vehicles crossing the Helmand River outside of Comp Payne with the use of U.S. Army Ribbon Bridge Boat.

Page 178, top: The day I caught an African Pigmy Hedgehog in the South Station gym while working out and brought it back to the tent to show the rest of the Marines.

Page 178, bottom: HM2 Bertucci takes fighting position while strapped with medical gear during hilltop overwatch in between three Afghan Villages.

Page 179, top: HM2 Bertucci is posing near Afghan villager home made of sticks and mud.

Page 179, bottom: The day the EOD vehicle (buffalo) hit an IED in a pinch point while entering a village. Several Casualties with minor injuries. Vehicle destroyed and parts scattered everywhere as they fell from the sky raining down on our dismounted patrol into the village.

Page 180, top: The back of the Ambulance LAV that HM2 Bertucci rode in and LCPL Orta drove. Display of stretchers, medical gear, and 240G Machine Gun.

Page 180, bottom: LCPL Hernandez (left) Molly (middle) HM2 Bertucci (right) standing outside of LAV scout hatch prepared to conduct dismounted patrol into Afghan Village.

Page 181, top: Random wild camel on mounted patrol rout in front of LAV25 as the sun sets.

Page 181, bottom: HM2 Bertucci is posing in front of Afghan village.

Page 182, top: Example of small Afghan homestead. The Afghan people lived without electricity or clean water inside of small huts constructed from mud and sticks.

Page 182, bottom: Very large explosion. Detonation of IED uncovered inside small Afghan village.

Page 183, top: LAV T.O.W. Equipped with Tube Launched Optically Tracked Wire Guided Missiles (near) and LAV 25 (far) on an active Afghan battlefield smothered in gunfire smoke.

Page 183, bottom: Vehicles staged during Operation Red Dawn 2 Battle (Pakistan Boarder) with burning enemy positions in the distance.

Pasge 184, top: Night Vision Scope of LAV 25 while it fires 25mm chain gun rounds at enemy position (Operation Red Dawn 2).

Page 184, bottom: Entire Bravo Company of LAVs mounted patrol *en route* to raid and destroy Taliban Pass.

Page 185, top: Marine on patrol among opium fields near small village with small curious Afghani children abroad begging for food and water.

Page 185, bottom: Marines posing with discovered marijuana fields that covered several miles of Afghani farming ground. (1)

Page 186, top: Marines posing with discovered marijuana fields that covered several miles of Afghani farming ground. (2)

Page 186, bottom: LCPL Trinh posing with village elder and children upon investigative questioning inside small village.

Page 187, top: Marines take fighting positions as firefights erupt during dismounted patrol.

Page 187, bottom: Marines on dismounted (foot) patrol inside a small village.

Page 188, top: Marines maintain dispersion while on foot patrol into unchartered territory in search of threats.

Page 188, bottom: HM2 Bertucci in fighting hole on hilltop outside of villages conducting firefights.

Page 189, top: HM2 Bertucci prepares for foot patrol into distant village.

Page 189, bottom: Pinch pointed route displaying another angle of destroyed EOD buffalo vehicle after striking a pressure plate IED while traveling into village.

Page 190, top: HM2 Bertucci covered in kicked up dust that poured into scout hatch of LAV 25 while on mounted patrol.

Page 190, bottom: Marines sweeping for explosives in rough channelizing terrain.

Page 191, top: LAV 25 on hilltop overlooking Helmand River.

Pasge 191, bottom: Larger living hut of Afghan Village. (view from Ambulance vehicle (Black- 8).

Page 192, top: HM2 Bertucci in water channel assisting Marines with supplies to other platoons inside of village.

Page 192, bottom: LAV coil with EOD attachments.

Page 193, top: LAV staging ramp inside of forward operating base.

A Poem Written By a Fellow Navy Corpsman

Veterans Past

"Warm, wet sand, stained red.
Quickly reminiscing the things never said.
Your eyes are closing; it will be okay.
Your pain will be over soon; blood flowed from your wounds like a typhoon.
How quickly from this life you were torn away.
It should be me, not you, I wish I was you. I should be DEAD.
Death comes like a thief in the night. Ironic; no stars in sight, this time it was day light.
The grim reaper has no remorse.
No matter if it wasn't on the course; if it's your time, it's your time.
He's stacking up bodies expecting us to feel FINE.
I want to hit pause, rewind, take another path and hit play.
Maybe you would be here to see another day.
I'm not that strong, this all feels wrong.
Things are not at all what they appear.
Continuing your legacy, has got the best of me, failing your dreams is my biggest fear.
Feels like the weight of the world on my back.
Constantly on guard, waiting for the attack.
Guilt was built like a boulder.
Just like your body, my heart became colder.

Now I'm a statue, just like you, no feelings or emotion.
Life is nothing but a cluster fuck; chaotic commotion.
You weren't the only one that met their demise.
I'm physically walking but I died inside, that's no surprise.
I'm haunted daily; woken up nightly by these vivid dreams.
This new life is old, drenched in sweat, BUT not what it seems.
Physically here but mentally not.
Eyes bloodshot; I feel my brain is starting to rot.
You have taken over and I'm the host.
Happiness is the thing that I miss the most.
I don't laugh and can't smile.
A grin is all I give, and that's going the extra mile.
This heartache and pain; I'm getting sick of it.
I'm truly tired of YOU and this FUCKIN shit!
I'm surrounded by happiness that I can't feel.
I begged and pleaded; nothing; so THIS is my final appeal.
You're gone we all know it; so leave me alone.
Goodbye my friend you're finally home.
Don't you remember your tribute?
It was stamped in time with a "Three-Volley gun salute".
You have no pain, you now have peace.
I've broken your chains, now it's time for your release.
Friends and family have moved on, but will never forget.
Holding you as you took your last breath, I don't regret.
You paid the ultimate sacrifice,
Not just for our country, but for my life.
Your looking down, I feel it, it was meant to be.
So please let me enjoy my family.
I raise my glass, no ice; just scotch.
Rest easy little brother, I've got the watch."

by HM2 Dominic Christofek USN
Iraq 2006 - 2007 / Afghanistan 2010

Veterans' Resources

There are countless resources for Veterans in the world. The number one answer to finding out what Veterans can do to better their lives, heal, and be compensated for their injuries is simple. It is ultimately up to the Veteran if he or she wishes to pursue the path to healing. It is highly recommended that friends and family encourage all Veterans to seek assistance and ask as many questions as possible to the ones placed in the position specifically to guide Veterans toward success. A brighter and more pleasant future awaits. These individuals are known as Veteran Service Officers (VSO). I urge you to seek out and make contact with your local VSO. I have had the pleasure of already helping thousands of veterans and their families as a District VSO. It brings me great joy to know that Veterans are receiving compensation payment for their injuries from their time in the military, education benefits, medical care, and just someone to talk to that can help get them on track toward a brighter future. Find your local VSO. They will help.

Special Thanks

I would like to extend my gratitude to those who have helped me in life. Those who have been there for me before, during, and after all that I have gone through and am still going through. Thank you for being there for me in my time of need, for listening to me when I needed someone to talk to. Thank you for showing me that there are good people in this world who also care for others. Thank you for encouraging me to complete this book and share it with others. Thank you for being a light in the darkness. I love you with all my heart and cherish you endlessly.

Ronald and Mary Ann Bertucci – Loving Father and Mother

Daniel Bertucci – Brother

James Allgor – Brother

Theresa and Louis Marchena

Christopher Daniele – Brother / War Comrade / Navy Corpsman

Chad, Cathy Jo, Andrew, Nick Beairsto – Family Friends / Life Mentors

CJ Calderon – Brother / War Comrade / Navy Corpsman

James and Bonnie Rice – Grandparents / Neighbors

Anthony Maple and the Maple Family – Family Friends

Gina Chandler – Sister / Supervisor

The King Family – Neighbors

The Athavale Family – Family Friends

Tom Willard – Naval Science Instructor / Family Friend / Life Mentor

Andrew Wall – Naval Science Instructor / Life Mentor

Jason and Caressa Marion – Friends and Supporters

All Members of the Armed Forces that I have had the pleasure of working with.

All other friends and family that I have had the pleasure of ever knowing and encountering…

About the Author

Nicholas Bertucci has always had love and compassion for helping and giving to others. His creativity helps him alleviate his pains from war. Through expressing his creativity, he combines it with his desire to serve others in the world and ensure they are happy, comfortable, and well taken care of. Nicholas wrote this book to continue to save lives, to share with others who would want to understand better what modern war is like and how it affects Veterans. He continues to care for others as he spreads awareness, knowledge, and love to everyone he encounters.

www.ingramcontent.com/pod-product-compliance
Lightning Source LLC
Chambersburg PA
CBHW072001290426
44109CB00018B/2093